BUILDING THE ATKINSON DIFFERENTIAL ENGINE

Written and illustrated by
Vincent R. Gingery

Published by
David J. Gingery Publishing LLC
P.O. Box 318
Rogersville, MO 65742

Printed in the USA

First Edition

First Printing 2000

Library Of Congress
Control Number 00-092581

International Standard Book Number
1878087-23-1

TABLE OF CONTENTS

Preface . . .

As I write this brief note, I have finished the work on this book and am readying it for the printer. I couldn't help but take these few minutes to reminisce. This has been a long project taking more than a year of constant work to complete. But even though I have written the book, the credit for the entire project must go to my Dad who spent many many hours working out the details and converting the Atkinson patent drawings into a workable engine long before a book was even thought of. Taking such a project and building it up from scratch is frustrating and I know that many times he would have liked to have just tossed this thing to one side and forgotten about it. But he was determined and his patience and determination is what made this project possible. I am fortunate and really appreciate the opportunity to work and learn from the person I most respect and admire.

As you begin this project, keep in mind that engine projects can be frustrating. It's the nature of the beast. But the rewards in the form of personal satisfaction are great. Often we get an engine built only to find it just won't run. And it often takes longer tinkering with the thing just to get it to run than it ever did to build it. This project is no exception to that rule. It's really a simple engine to build, but after it's built you will discover it requires a great deal of patience and many hours of assisted run - time to achieve success. Assisted run time means connecting the flywheel of the engine to an electric motor by means of a "V" belt. The engine is powered by the electric motor for a period of time until the rings seat and it is able to run on its own. The main issue with this engine is compression. The reason appears to be that instead of the pistons compressing against a solid head they are compressing against each other. And they are doing so in the same cylinder space. So two pistons and two sets of rings must seat in the same cylinder or there will be compression leaks around one or both sets of rings. And without good compression an engine just won't run. It would seem that how well you do on your cylinder and how well the pistons and rings fit the cylinder will go a long way in determining how quickly your engine runs. Even so, expect to spend several hours of break-in time before this engine finally runs under its own power.

Introduction . . .

I guess it was only natural that after building a version of the "Atkinson Cycle Engine" a couple of years ago we would become interested in another one of the James Atkinson engines. This one being his Differential engine and the subject of this book. The engine is named for the peculiar differential motion of its 2 pistons. The purpose of this book will be to first cover a bit of history concerning James Atkinson. Then we will discuss the operating principles of his Differential engine and finally we will move on to the actual construction of the engine. Much of the following historical information comes from a book published in 1900 titled, "Gas, Oil, and Air Engines" by Bryan Donkin and also a book titled, "Internal Fire" written in 1976 by Lyle Cummins. Construction details were derived from the original patent drawings and text of 1886 taken from the book, " IC Engines Vol. 1" published by and available from Lindsay publications.

James Atkinson was an amazing man. While studying his patent drawings and building two of his engines it seems hard to believe that this is information and discussion from over 100 years ago. The rapidly growing industrial age that Atkinson lived in was similar to the technology age we live in where the dominating presence is Microsoft. But in Atkinsons time the dominating presence was Nicolaus Otto and the amazing technology was the Otto 4-cycle engine which had been in production since 1876. The Otto engines were in high demand and practically no other engine type was being constructed. They were the favored technology of the day and were being used to power elevators, water pumps, entire factories and were even producing power for another amazing phenomenon, the electric light bulb. The Otto engine touched every aspect of life at the time and Otto had a strong hold on the engine market. During the years 1876 to 1890 Gasmotoren-Fabrick Duetz and its licensee protected their position as the worlds sole supplier of the Otto engine. They ruled with an iron hand and through litigation and threats challenged all builders who attempted to enter the engine market or infringe on the Otto patents. The issues involved not only the 4-stroke engine, but every other compression engine as well. Nothing went unchallenged and competition was virtually eliminated.

No one could deny the fact that the Otto engine had its defects. Experiments had proven that even with the best Otto engine less than 25% of the fuel consumed actually ended up producing power with the other 75% of fuel being wasted. Defective expansion was given as the chief cause of this. So the challenge occupying the engineers of the day was how to increase the length of the piston stroke and admit more combustible charge yet at the same time allow a greater scope for the expansion of the gases. An ingenious solution to the problem came in 1885 when James Atkinson introduced his Differential engine at the inventions exhibition in London.

Mr. Atkinson, born in England in 1849 had started his career working on steam engines as an apprentice for a marine engineering company located in Jarrow England. Later, as a successful design engineer he filed his first patent in 1879 for a hot tube igniter. During the following 3 years he filed patents for a couple of Robson type engines. Then in 1883 a company called "The British Gas Engine & Engineering Co., Ltd." was formed to build the Robson type engines with Atkinson being appointed managing director. But it is really not until 1885 when he filed his patent application for the Differential engine and then again in 1886 when he introduced his Cycle engine that we become interested in him.

Even though the Differential shown in figure 1 and the Cycle engine shown in figure 2 look very different, the operating principle in each is the same, just carried out in a different manner. The unique thing about both of these engines besides their unusual appearance is their ability to complete all 4 strokes (intake, compression, power & exhaust) in a single revolution of the crank with the whole operation taking place in one cylinder. Atkinson's opinion was that the two main sources of wasted energy were the exhaust and the water jacket. He attempted to reduce these losses by arranging the connection between the piston and the crank so as to give different lengths of stroke. He reasoned that if the piston travels more quickly there is less time for the heat to be carried off by the jacket. And if a longer expansion stroke is obtained, the heat and pressure of the gases have more time to act in doing useful work on the piston before the exhaust opens. These thoughts were the basis for his designs of both the Differential and the Cycle engines. Lyle Cummins says in his book, "Internal Fire" that Atkinson, through his complex linkage arrangements was able to achieve what no other designer has since achieved: an expansion stroke double that of the intake.

The Differential engine was offered for sale in the 1, 4, and 8 horsepower sizes for a couple of years. I was unable to find any mention of how many of these engines actually sold, but I suspect not very many. The engine was interesting to look at and the concept was good, but the levers, links and connecting rods turned out to be a problem for these engines and as in all technology improvements were made. Atkinson soon abandoned the Differential engine to work on the development and production of the more practical and successful Cycle engine design.

Figure 1 *The Gingery version of the "Atkinson Differential" engine.*

The Cycle engine had a short run success between 1886 and 1890 with well over 1000 units being sold. But when the Otto patent expired in 1890 most manufacturers took advantage of the opportunity and began producing the Otto engine in large numbers. The market was flooded and competition was fierce. The selling price of the Otto-type 4-stroke engines plummeted driving many out of business. Atkinson was no exception. He made one last attempt to revitalize his company with the introduction of the "Utilite", a two cycle engine, but the attempt failed. The British Gas Engine company went out of business in 1893.

Soon after, Mr. Atkinson went to work for Crossely Brothers., Ltd. as chief engineer where he continued to make improvements to the gas engine. He retired in 1912 and died two years later.

Figure 2 *The Gingery / Lewis version of the "Atkinson Cycle" engine.*

In Atkinson's own words . . .

This is a fascinating engine and I thought it might be fitting to give Mr. Atkinson himself the first shot at explaining his "Differential" engine. And, he is able to do so more than 86 years after his death through his patent application. Later I will fill in the blanks through the use of a wooden mockup, but for now have a look at the next 6 pages which contain the reproduced text and drawings from the original patent dated February 16, 1886. This patent information comes from Lindsay Publication's book, "IC Engines Vol. 1". This is just one of 14 other patent applications for engines dating from 1881 through 1890 that can be found in his book! And it's not the first time someone has built an engine from the patents found in the Lindsay books. Jim Lewis got the necessary information he needed to build his "Linford" engine from "IC Engines Vol. 2". For more information on the patent volumes and other books write to:

Lindsay Publications
P.O. Box 538
Bradley, IL 60915-0538
Phone 815 935 5353
Or visit the Lindsay web page at
http://www.lindsaybks.com

As you read through Atkinson's patent application you will discover how the patent process is really quite fascinating. It was the job of the applicant and his attorneys to provide just enough information to give a general idea of what the applicant had to offer. But you didn't want to give away all the information so there was always a little bit left to the imagination. Even so, the important information we needed to build our engine was found here. First by determining the size engine we thought we would like to build. Then by enlarging the patent drawings on a copy machine until they reached that size. In this instance, the original drawings were first enlarged 200% and then an additional 110%. Then the enlarged drawings were used as a guide to construct a wooden mockup of the engine. The wooden mockup was instrumental in giving us a 3 dimensional view of the project and also helped us determine if the engine would really be practical to build. Later the information derived from the wooden mockup was used to construct wooden patterns of the individual parts. The patterns were rammed up in sand molds and aluminum castings of the individual parts were poured.

UNITED STATES PATENT OFFICE.

JAMES ATKINSON, OF HAMPSTEAD, COUNTY OF MIDDLESEX, ENGLAND.

GAS-ENGINE

SPECIFICATION forming part of Letters Patent No. 336,505, dated February 16, 1886

Application filed May 20, 1885. Serial No. 166,131. (No model.) Patented in England February 28, 1885 No. 2712; in France May 21, 1885, No. 169,071, and in Belgium May 30, 1885 No. 69,074.

To all whom it may concern

Be it known that I, JAMES ATKINSON, a subject of the Queen of England, residing at Hampstead, in the county of Middlesex and Kingdom of England, have invented new and useful Improvements in Gas-Engines, (for which I have applied for Letters Patent in Great Britain, No. 2,712, bearing date February 28, 1885,) of which the following is a specification.

This invention relates to and consists in a novel arrangement of parts forming a gas-engine of the compression type.

As hitherto constructed compression gas engines may be broadly divided into two types. In one type the same piston (or pistons) is used alternately as a pump and as a working piston, requiring more than one revolution to complete the cycle of operations. In the other type an independent pump is used.

In this invention I employ a cylinder open at each end, fitted with two pistons, which have a peculiar differential motion as regards each other, which causes one of them to operate as a working-piston and the other as a pump-piston, in such a manner that in one cylinder and during one revolution only of the main shaft, they draw in the combustible mixture, compress it, ignite it, develop its expansive force, and expel it.

In the accompanying illustrations a suitable arrangement of this invention is shown. Figure 1 diagrammatically shows the positions of these pistons and of the crank-pin at the termination of the exhaust; Fig. 2, the capacity of the charge; Fig. 3, the space for compression of the charge, and Fig. 4 the capacity for the expansion of the charge after ignition, being four different positions of one revolution of the main crank-shaft "C", "A" representing the pumping piston, and "B" the working piston. Fig. 5 is a vertical front elevation of the complete engine. Fig. 6 is a front section through the cylinder "D". Fig. 7 is a vertical section at right angles with Fig. 5.

In this arrangement I use a horizontal Cylinder, "D", fixed near the base of the engine and fitted with the pump-piston "A" and the working-piston "B". Each of these pistons is connected by a rod or link to the lower end of a beam, the pumping piston "A" being connected to the beam "E", and the working-piston "B" to the beam "F". These beams vibrate on centers fixed in the main framing of the engine, above the open mouths of the cylinder, and are extended above these centers, where their upper ends are coupled by short connecting rods to one crank-pin, "G", in a crank-disk, "H", fixed on the main shaft "C", which works between the upper ends of the beams "E" and "F". The connecting-rods are made very short, (say one and one-half time the throw of the crank-pin.) The angularity of the rods and beams causes the pistons to have a differential motion as regards each other. They are quite close together when at one end of the cylinder, (hereinafter called the "preparing end,") as shown in Fig. 1. The results of the last working charge having been driven through the self-acting exhaust-valve "I" into the exhaust-pipe as the crank-pin "G" revolves, the working-piston "B" being, at the preparing end of the cylinder, moves very slowly toward the other end, (hereinafter called the "working end.") At the same time the pumping-piston "A" travels very rapidly, so that the pistons leave a considerable space between them, into which combustible mixture is drawn through the self-acting suction-valve "K", this position being shown in Fig.2. The working-piston "B" now begins to travel faster than the pumping -piston "A", covering the port to the admission-valve "K", until they arrive at the termination of their strokes at the working end of the cylinder "D", when the pistons are closer together and have compressed the combustible mixture to a suitable pressure, as Shown in Fig. 3. At about this time the pumping-piston "A" uncovers the ignition-port, admitting some of the compressed charge to the inside of the small tube "L", which projects from the cylinder "D". This tube "L" is kept red-hot by means of in external flame and fires the combustible mixture admitted to its interior, thus igniting the charge. The piston "A" now remains almost stationary, while the working-piston "B" is rapidly passing from, the working-end to the preparing end of the cylinder "D", thus giving a very quick expansion. At the termination of this expansion the position of the pistons and crank-pin areas are as shown in Fig. 4. The pumping-piston "A" now commences to overtake the working-piston "B", driving out the products of combustion through the exhaust-valve "I", the opening to the valve "K" being now uncovered by the piston "B", when the pistons again arrive at the positions shown in Fig. 1, thus completing the cycle of operations during one revolution of the crank-shaft.

I prefer to make the space between the pistons, into which the ignited gases are expanded, as shown in Fig. 4, greater than the space into which the combustible mixture is drawn, as shown in Fig. 2, thus carrying the expansion to a further extent than in gas-engines as hitherto constructed. The relative proportions of these spaces and the capacities between the pistons when at the extremes of their strokes are obtained by suitably proportioning the levers, rods, and working centers, the short connecting-rods between the crank pin and the upper ends of the beams being chiefly the cause of the great difference in the relative speeds of the pistons, and the level of the main shaft as regards the upper ends of the beams principally determines the relative spaces for the combustible charge and for expansion. The admission-port and the exhaust port are at the preparing end of the cylinder, and the admission and exhaust are chiefly controlled by the working-piston covering or uncovering their ports. They have self-acting valves to prevent the passage of air, gas, or the exhaust passing in the wrong direction; also the exhaust-pipe may be provided with a self-acting-valve admitting air to the exhaust-pipe, so as to prevent the momentum of the exhaust drawing unignited gas and air through the engine into the exhaust-pipe; but it might be allowed to draw a little air only through the very small space between the pistons at the termination of their strokes when in the position shown in Fig. 1.

The ignition may be caused by other means than by the ignition-tube "L", previously mentioned. This, however, is a very simple and certain method, and has the great advantage of being very accurately timed by the pumping-piston passing the ignition-port

By the inspection of Fig. 3 it will be observed that the working-piston "B", completes its stroke (when at the working end of the cylinder) slightly before the pumping-piston "A" completes its stroke. This insures the ignition-port, being fully open at the right time and causes the pumping-piston "A" to remain practically stationary until the working-piston "B" has traveled far enough to have very materially reduced the high initial pressure.

The cylinder may be placed vertically or inclined, and more than one cylinder may be connected, so as to form a combined engine.

The working-gear may be differently arranged to suit different circumstances as, for instance, the beams may vibrate from one end, with the cylinder placed between the shaft and these beam-centers, or the shaft between the cylinder and the beam-centers, the essential point being to get the peculiar differential motion of the two pistons. No working gear is necessary, as all the valves may be self-acting. It may, however, be desirable to control the gas-supply so that no gas may pass into the engine, excepting at the desired time, also for the purpose of governing the engine.

The engine may be arranged to run in either direction, as started; but if required to be reversible, so as to work both ways, the spaces for drawing in the combustible charge and for expansion should be made equal, or about equal.

It will be observed that the preparing end of the cylinder works cooler than the working end, and that only a very small portion of the cylinder requires to be water-jacketed; also the expansion of the ignited charge is completed two to four times as rapidly as in other engines, (making the same number of revolutions of the crank-shaft in the same time,) consequently a much greater percentage of the heat developed by the ignition of the charge is absorbed in transmitting power to the crank-shaft and less wasted in being transmitted to the water in the jacket. It is well known that in gas-engines as hitherto constructed one-half and over of the total number of units of heat in the gas burned in the engine is wasted in this manner, and by developing the power in the manner described in my invention at such a greatly-increased speed a most important economy is effected. By my invention, also, the working parts of the engine are very much simplified and rendered more durable. There is no slide-valve, which is such a constant source of trouble in most other engines. The suction and exhaust valves are simple self-acting lift-valves and are not subject to any material pressure. Slight leakage past them has very little effect upon the working of the engine. There are no delicate or intricate parts in connection with the engine, and the cost of production is materially reduced.

What I claim, and desire to secure by Letters Patent, is-

1. In a gas-engine, the combination, with the open-ended cylinder, of the two pistons and beams connected therewith centrally pivoted and connected to the same crank at their outer ends, as set forth.

2. In a gas-engine, the combination, with the open-ended cylinder, the two pistons working therein with different relative velocities, but each Making the complete stroke in the same time, of the centrally-pivoted beams connected to said pistons and to the same crank at their outer ends, as set forth.

3. In a gas-engine, the combination, with the open-ended cylinder and pistons moving therein at the differential velocities, herein set forth, and beams connected to said pistons and at one end and to a single crank at the other, of an ignition-tube located and adapted to operate as described.

4. In a gas-engine, the combination, with the cylinder and differentially-moving pistons connected to beams working from the same crank, of the admission-valve constructed and applied to be automatically operated by the movement of the pistons, and an exhaust valve also constructed and applied to be operated by the pistons.

5. In a gas-engine, the pistons A B, working in a cylinder which is open at both ends, completing their strokes at the same times, connected directly with the crank-pin of the driving-shaft, so as to maintain a certain position in relation to the crank-pin at every part of its revolution and in such a manner as to work with a differential rate of motion to allow the admission, compression, ignition, expansion, and expulsion of the working-fluid to be performed automatically between said pistons, as set forth.

JAMES ATKINSON

Witnesses:

Arther Coleman,
George Edward Priddle.

(No Model.) 3 Sheets—Sheet 1.

J. ATKINSON.

GAS ENGINE.

No. 336,505. Patented Feb. 16, 1886.

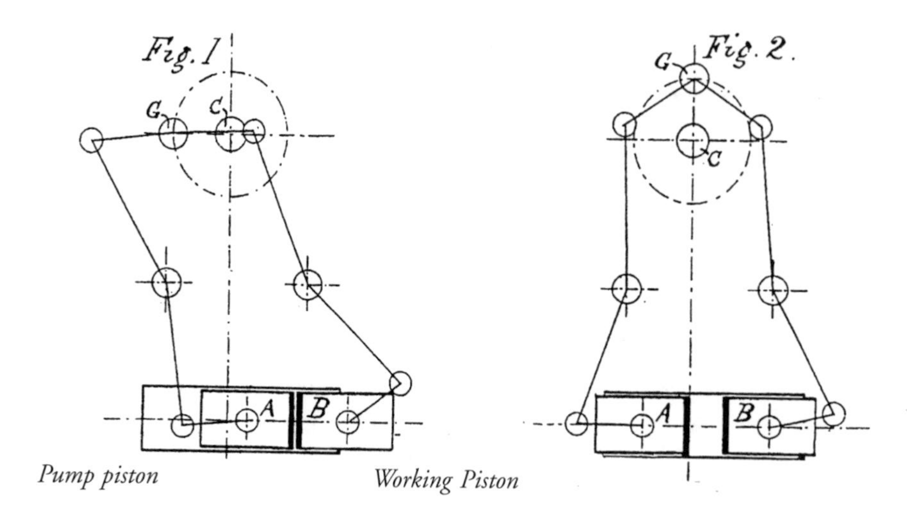

Postitions at end of Exhaust

Postitions of receiving charge

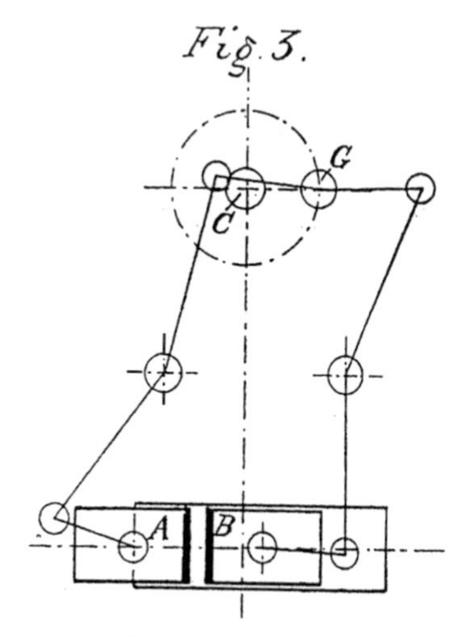

Postitions at compression

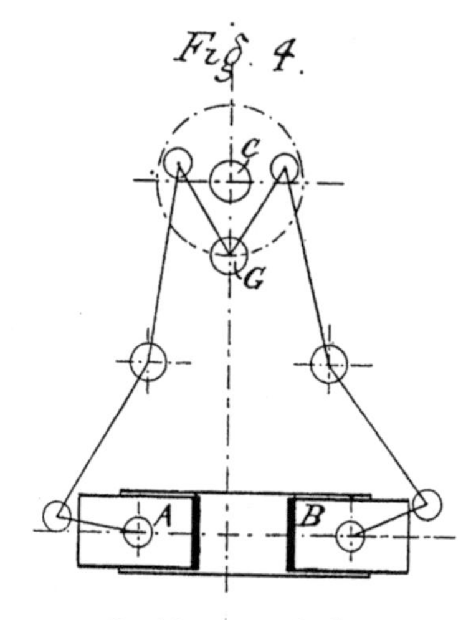

Postitions at end of Expansion

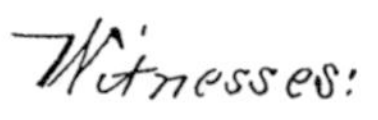

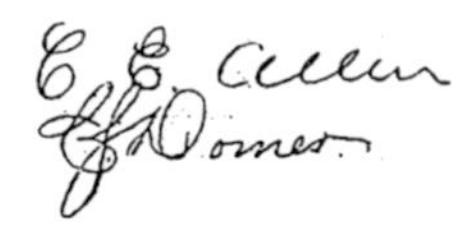

Inventor:-

Jas. Atkinson

By atty J.M. Kalb

(No Model.)

3 Sheets—Sheet 2.

J. ATKINSON.

GAS ENGINE.

No. 336,505.

Patented Feb. 16, 1886.

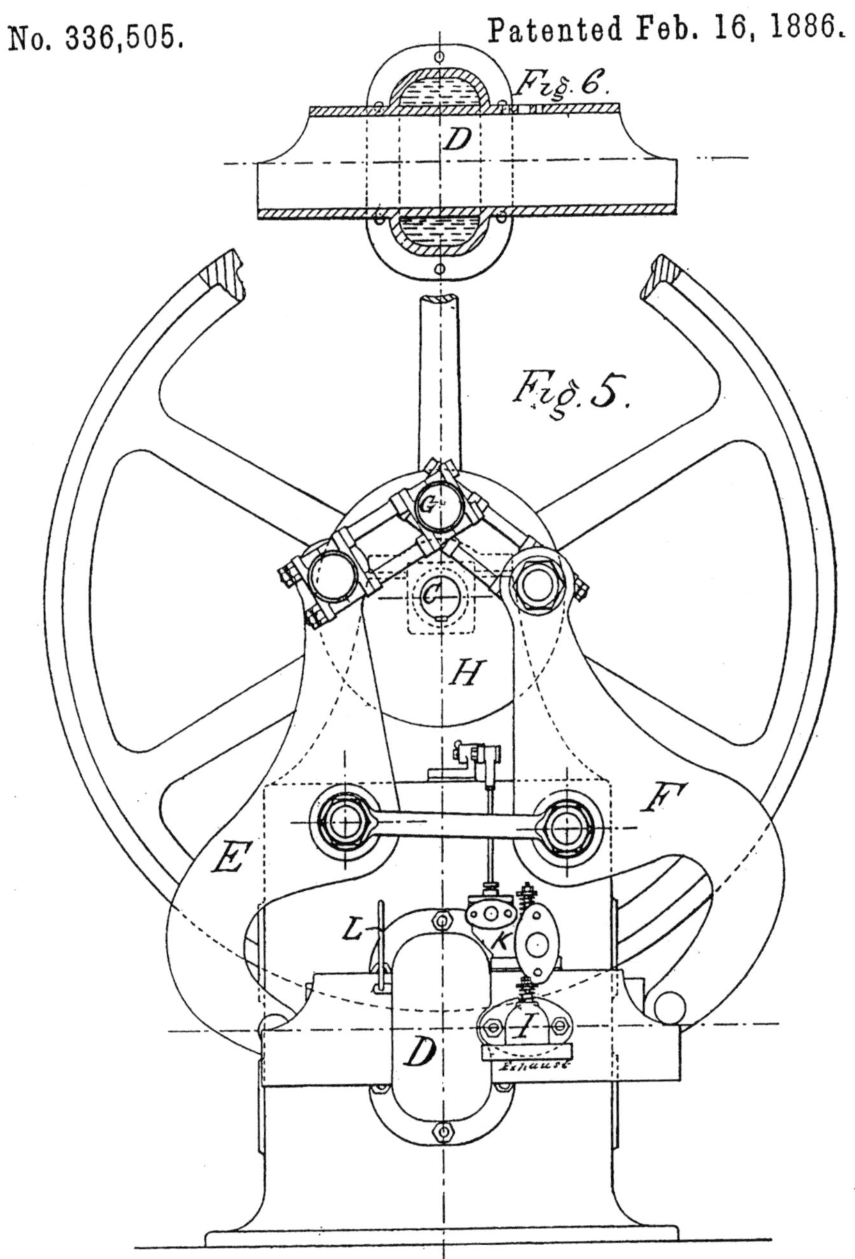

Witnesses:-
B.C. Allen
C.J. Domer

Inventor:
Jas. Atkinson
By atty J.M. Kalb

No. 336,505. Patented Feb. 16, 1886.

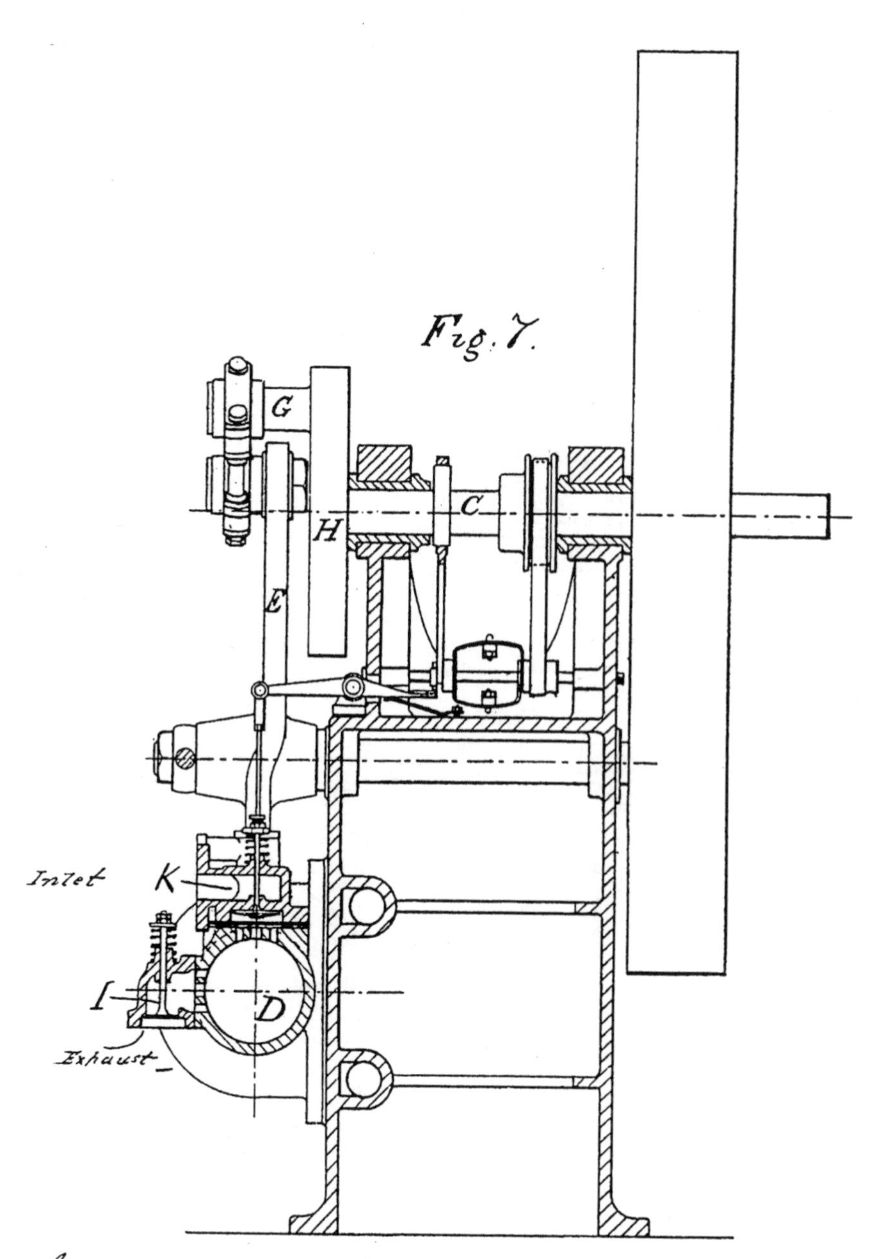

Witnesses:

C. E. Allen

C. J. Dorner

Inventor:-

Jas. Atkinson

By atty J. N. Kalb

A few words from Dave Gingery . . .

The most common request I get when people ferret me out is for help in developing a project. I am always sorry to decline but I simply don't have time available for the work. Nor is it possible to give advance information on projects Vince may be developing for publication. That is why I have urged Vince to offer these guidelines on developing your own projects. And the development of the Atkinson differential engine project should prove helpful for you to see how a project evolves from conception to delivery.

While the inspiration or project concept may come from many sources it usually begins with seeing something others have made. That was the case for us with the differential engine. And though some plans were available we wanted to more accurately replicate the original work of James Atkinson. And so with patent drawings in hand we proceeded to build a wooden model to work out dimensions and develop design concepts that would later adapt to metal parts for the running engine. This is the basic method I have used to develop all of my machine and engine projects and it is most efficient and effective.

Patent drawings are intended to demonstrate to patent officials that an idea is original, that it works and has a practical value. But they are not intended to instruct anyone on how to build a working model since the inventor hopes to reserve that right to himself for the period of time covered by the patent. So while there is sufficient detail to fabricate some representation of the device you may be certain that details will be missing or obscured. Your challenge is to discover the secret. And if a patent was granted you may be certain that the secret is revealed in the record of drawings and claims.

The very first requirement in any project is to see it from every angle so plans and drawings are very useful. But even a fist full of drawings may not be adequate for something as intricate as an engine. In these pages Vince will show you how we took patent drawings, enlarged them to practical size and used them to produce a working wooden model from which we developed the project. Thus we were able to actually see the mechanisms in motion and establish critical dimensions for pivot points, establish locations for valve and ignition ports and gain a sense of proportion that enabled us to trim it out for authentic period appearance.

Now even if you have no intention of building an Atkinson differential engine I urge you to build this wooden mock-up and use it to demonstrate to yourself and others how the engine works and how the model is useful in the development of the project. The cost will be slight and the reward for effort spent will be great. By doing that you will learn these important skills:

1. How to interpret a plan drawing or sketch.

2. How to combine plan drawings of two or more views into an angular perspective.

3. How to transfer dimensions from sketch to workable materials.

4. How to shape the materials to usable parts.

5. How to assemble the parts into a working model.

By using wood, cardboard, paper, plastic and other cheap materials you create a working model you can use to prove the concept. As you solve the relatively easy problems of producing the model parts you lay the foundation for solving the more demanding problems of producing the actual metal parts later. And you will have a demonstration model that enables you to show others how your machine works in a clear and convincing way.

Now I leave the serious work to Vince and the pleasure of a rewarding project to you.

Dave Gingery

Building a wooden mock up. . .

Inevitably, the first question someone will ask when they see your engine is, "how does it run?" And probably the best tool for explaining this will be the wooden mock-up.

It's a simple project, and very few woodworking skills are required to construct it. We created our wooden mock-up from information found in the patent drawings and you can too. First we determined the scale to build and then enlarged the drawings on a copy machine to achieve that size. Measurements were taken from the enlargements for the front, rear and side panels and the oscillating arms, cylinder and water jacket. The dimensions were used to lay out patterns for the individual parts.

Instead of using the patent drawings you might consider using the same information found in the pattern section of this book. When it's finished you will have a 3-dimensional view of the main body of the engine which really helps put things in perspective as you build the real thing. Not much material is required other than a bit of 1/4" plywood, a couple of short lengths of 1/2" and 1" wooden dowel rod, a bit of hardware and some wood glue.

The cylinder is made of 1-1/4" pvc pipe and has a 1/2" wide slot cut along the entire length of its front. This gives us the ability to view the pistons as they travel.

The jacket that holds the cylinder in position was formed from a short piece of 2X4 lumber. As you can see in the photo a hole was drilled through the jacket for the cylinder and a nice radius was turned on the front corners of the jacket. The assembly is attached to the front panel with wire brads and glue.

The two wooden pistons contained in the cylinder are made from 1" dowel rod. They required a bit of innovation because that information was not found in the patent drawings.

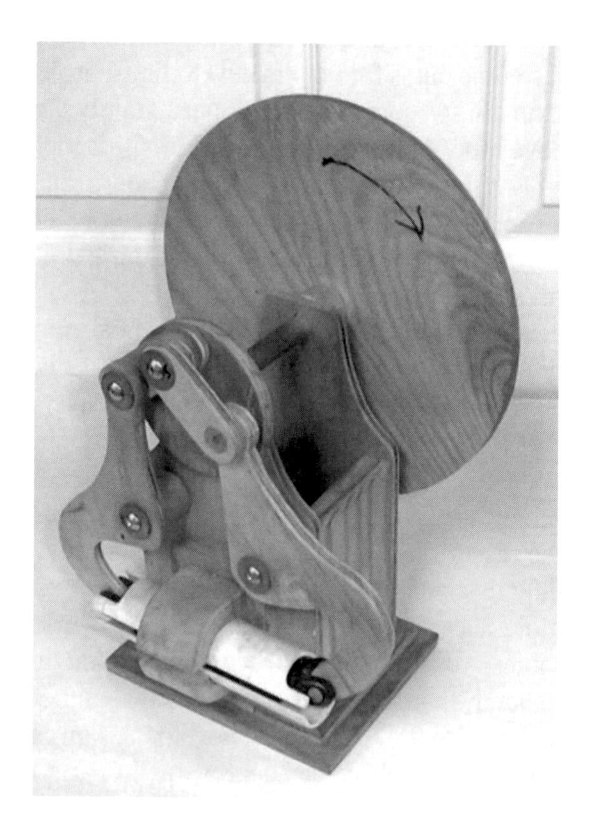

Figure 3 *The wooden mock up*

A piston rod connects each piston to an oscillating arm. The piston rods also required a bit of innovation because that information was lacking.

Each oscillating arm pivots on a 1/2" dowel rod shaft that runs through the main frame. The upper ends of the oscillating arms are connected to the crank disk by means of short connecting links. Screws and washers are used to prevent the arms and links from rotating off their shafts.

A couple of coats of polyurethane or varnish and your demonstration model is ready to go.

The 4 photos of the wooden mock-up on the next two pages will help explain the basic operation of the engine. A visual of the valves, spark plug and points are not pictured yet, so you will have to use a bit of imagination. The main value of this demonstration is to show the position of the pistons at each of the 4 cycles. Later we will go through it again in greater detail with photos and drawings showing the spark plug, ignition points and valves in relation to each cycle.

Just as Atkinson did, we have referred to the piston at the right side of the cylinder as the working piston and the piston at the left side of the cylinder as the pump piston.

Atkinson also refers to the right side of the cylinder as the preparing end and the left side of the cylinder as the working end.

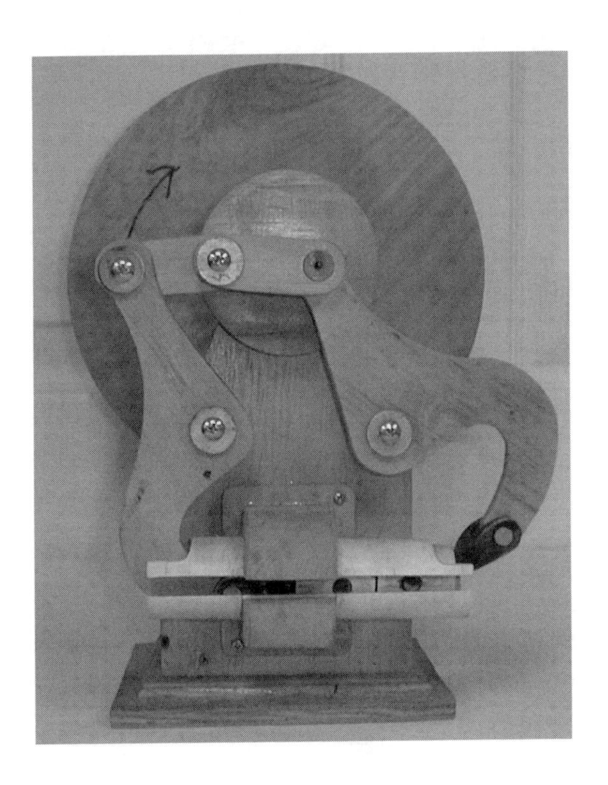

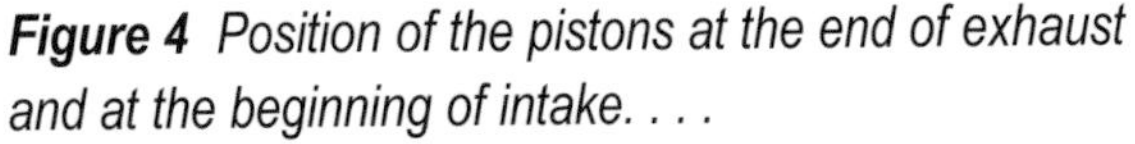

Figure 4 *Position of the pistons at the end of exhaust and at the beginning of intake. . . .*
The products of combustion have just been driven out of the cylinder. The clearance between the 2 pistons is reduced to its smallest point. In fact they almost touch each other. As the flywheel continues to rotate to the right in a clockwise direction, the pump piston will begin to move rapidly to the left. The working piston will follow at a slower rate and the space between the 2 pistons will increase. It is at about this time the self activating intake valve opens and the fuel charge enters between the pistons.

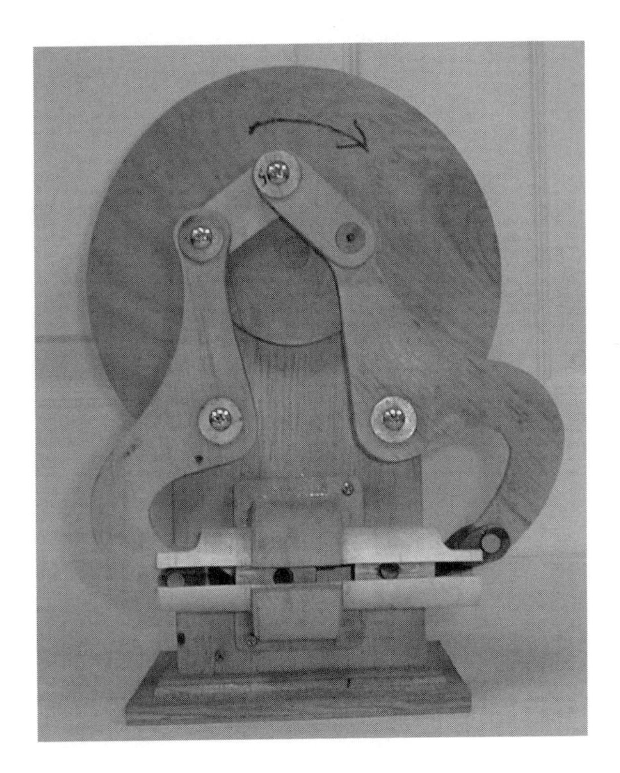

Figure 5 *Position of the pistons at the end of intake and the beginning of compression. . . . .*
The intake valve has just closed. The pistons are about 5/8" apart. As the flywheel continues to rotate, both pistons are continuing to move to the left with the working piston rapidly overtaking the pump piston and compressing the mixture.

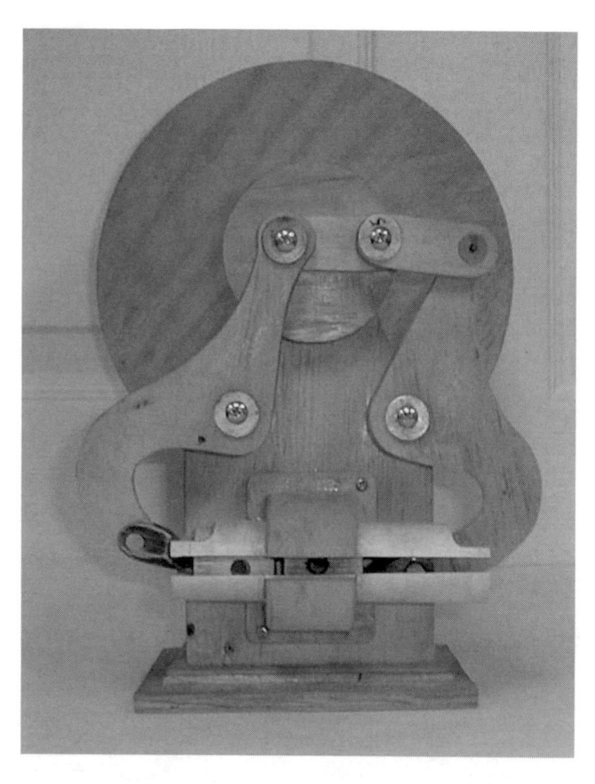

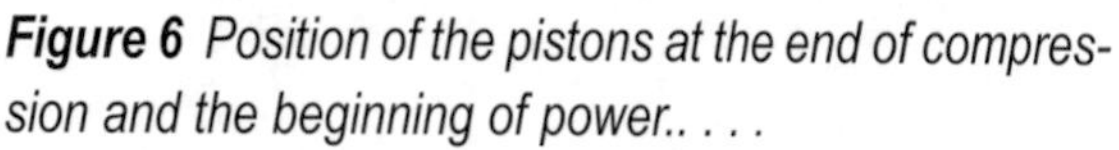

Figure 6 *Position of the pistons at the end of compression and the beginning of power.. . . .*
The spark plug hole in the cylinder wall has just been uncovered. And it is at this time the points break contact producing a spark that ignites the compressed mixture. At the point of ignition the clearance between the 2 pistons is about 1/8" and the force of the igniting mixture causes the working piston to move rapidly to the extreme right of the cylinder.

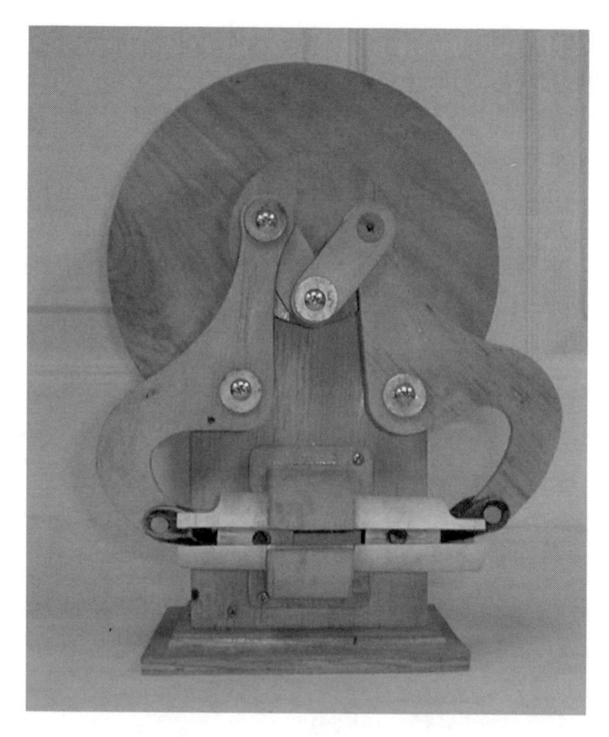

Figure 7 *Position of the pistons at the end of power and the beginning of exhaust.. . . . .*
The power stroke has just ended leaving the working piston at the extreme right of the cylinder. At this time, the space between the pistons is at its furthest point. The pump piston is following the working piston at a slower rate and will soon catch up. As it continues to move to the right, the exhaust hole is uncovered and the self activating exhaust valve opens to let the spent mixture escape into the atmosphere.

Before starting . . .

We have chosen to use aluminum castings for several of the parts in this project. Those items being, the flywheel, base, oscillating arms, piston rods, main frame side, front and rear panels. If you don't have casting ability in your shop, don't be concerned. Just because we use castings does not mean you must. The items mentioned above could be machined from solid blocks of steel or aluminum. To do so, use the same information found in the pattern making section, but instead of making the items from wood you would be making them from metal. No doubt you will create a few more metal chips in the process of machining, and it'll take more time, but the same or better results can be achieved. In fact, many who have built our engine projects in the past have chosen to machine the parts rather than cast them and in the process have achieved excellent results.

I may pour a few extra sets of castings for this engine and offer them for sale. For more information look on page 112 under the heading castings.

If you don't have casting ability in your shop you might consider building a small foundry. It's really not that hard nor is it expensive. And you will soon find it's a lot easier to carve your parts out of wood than metal. Several good books on the subject are available and at least one company manufactures a small melting furnace for the home shop at a reasonable cost. (See the information list at the back of book for more information.)

The next few pages of the book will cover an introduction to the pattern making, molding and casting procedures needed for the project in this book. Hopefully, this will help those of you who have not experienced the process gain a bit of insight into just how fun and easy it is. Metal casting really does open a whole new world of possibilities in the home shop.

After the parts are cast we'll begin construction of the engine which I think you'll find surprisingly simple. The valves are self activating so no complicated linkage is needed. Since all 4 cycles occur in a single revolution of the crank, no timing gears are required. And the carburetor is a simple straight tube type with a needle valve used to control the fuel mixture.

Even though I have characterized the engine as fairly simple to build, reasonable precision is required in making the parts. You'll need a lathe and milling machine as well as a variety of other tools, fixtures and accessories to aid in construction, but we'll discuss this in more detail later before actual construction begins.

Safety . . .

As in all projects, keep safety foremost in your mind as you proceed. There are dangers in this project both hidden and obvious. Take the time to think each step through before you begin. It's impossible to point out all the dangers. So you need to be alert at all times! No doubt you have heard all the warnings before, but be alert anyway! Wear eye protection! Revolving drills, saw blades, mill cutters, flywheels etc., pose a constant danger. A moment of carelessness around a machine can cause the loss of a finger or hand or worse. Long hair, jewelry and loose fitting clothing can get caught in a machine and cause serious injury. As mentioned earlier, there are other dangers too. Use common sense and understand all of your machinery making sure it is in good working order. Expect the unexpected! Keep your floor swept clean and post clear warnings as a reminder to yourself and others.

Figure 8 *The castings*

An overview of casting . . .

For those who may be unfamiliar with the casting process a brief description follows. If you decide you want to learn more, check out some of the great books on the subject available from the suppliers listed at the back of the book.

All of the castings in this project were produced using sand molds. Simply put, the procedure is to embed a pattern in sand in a 2-part mold. When the pattern is removed from the sand its impression is left in the mold. Then molten metal is poured into the mold impression. It solidifies and takes the shape of the pattern impression left in the mold.

For the whole process to work, the sand must be able to hold its shape and bond together after the pattern is removed. To gain this characteristic, the sand is mixed with an adhesive agent. In Green-Sand molding, the adhesive agent is clay and water. Another type of molding sand available commercially is a petroleum based oil bonding sand. Either green-sand or petroleum based sand will work fine for this project.

The molds in this project are formed in 2-part flasks. Flasks are wooden boxes with no bottoms or tops. They fit together with pins at each end so the halves can be aligned in perfect register each time they are put together.

The bottom half of the flask is called the drag and the top halve of the flask is called the cope.

A typical molding process consists of placing a pattern on a flat board called the molding board. The drag half of the flask is inverted over the pattern. Parting compound is sprinkled over the pattern and bottom board to prevent the sand from sticking to them. The flask is rammed full of molding sand and struck off level. Next, the mold is vented using a length of light gauge wire which is inserted into the mold and almost to the pattern in a few places. This provides an escape for steam created by the molten metal as it enters the mold cavity. Next, a bottom board is rubbed in over the sand, the drag is rolled over and the molding board set aside. The bedded pattern is now visible.

Next, the cope half is set in place on top of the drag. Dust the pattern and mold surface with parting compound. Then the sprue pin is set in place. The cope is rammed full of sand and struck off level. The sprue pin is removed and the cope is carefully lifted straight up and set on edge behind the drag. All loose dust is blown away and the edges of the mold are strengthened by swabbing with a soft brush dipped in water. After cleanup the pattern is rapped lightly in all directions to free it from the sand, and then it is removed from the mold. Screw eyes screwed into the pattern can be used to aid in its removal. After the pattern is removed, a gate is cut between the sprue opening and the pattern cavity. *A gate (sometimes called a runner) can best be described as a neat little trench cut in the mold surface to open up a path for the molten metal to flow into the mold cavity. Then the mold cavity is cleaned up by removing loose particles and strengthening or repairing damaged edges. When cleanup is done, the mold is re-closed and ready to pour.

The above operation was described in its most basic terms. There are variations to the procedure depending on what one is casting. For instance, some of the patterns in this

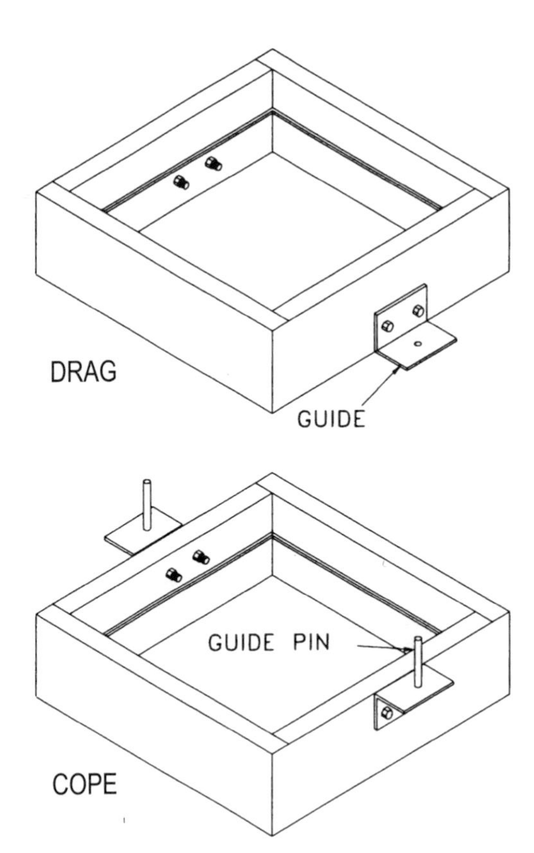

Figure 9 *Typical homemade flask*

project will be split. This means that the pattern is made in 2 halves and is split along the natural parting line. The natural parting line is that line which divides the cope and the drag. Split patterns are held in perfect register by pins called pattern dowels. The pattern dowels are installed so the pattern can only be put together in one way. The molding procedure for a split pattern would be to ram up one half of the pattern in the drag. After the drag is rolled over, the pattern halves are put together and the cope is rammed up as before. When the mold is separated, there will be a pattern half in both the cope and the drag.

* A gate is the path cut between the sprue cavity and the mold cavity. Sometimes the gate is part of the pattern design which eliminates the need to manually cut it in.

An overview of patternmaking . . .

The patterns in this project require only basic woodworking skills to make. They are constructed of white pine, 1/4" plywood and wooden dowel rod. Carpenters glue and/or wire brads are used to join pattern parts. Polyester auto body putty can be used for filling imperfections in the pattern surface and for wiping fillets on inside corners of the pattern. After completion, all patterns are sanded smooth and coated with varnish, paint or polyurethane to prevent moisture absorption.

The base, flywheel, oscillating arms and piston rod are split patterns and the front, rear and side panels are single patterns.

When making patterns there are a few common things that you will need to know. They are as follows:

First of all, you need to be able to remove the pattern from the mold without damaging the sand impression. For this reason, we give the vertical edges of the pattern sloped edges. This is also referred to as giving the pattern draft. Pattern draft can be as little as 1 degree for an experienced molder to as much as 3-5 degrees for one that is inexperienced. In the pattern drawings that follow, draft direction will be shown with an arrow.

Second, when making the patterns, outside edges must be given a radius and inside corners given a fillet. The object is to eliminate any sharp edges or angles that could weaken the mold.

And you need to realize that molten metal shrinks upon solidification and that castings are often machined to their final size. We have made allowances in our patterns for shrinkage, but just so you will know, aluminum usually shrinks approximately 3/16" per foot in all directions. So a pattern 12-3/16" long will produce a casting 12" long.

The most important thing to keep in mind when making patterns is to allow plenty or draft and make sure pattern surfaces are smooth.

Constructing the base pattern . . .

We originally made the base without legs, but later decided to add them in an effort to make it easier to level the engine. The added legs make this a split pattern. The main base of the pattern being in the drag half of the flask and the legs being in the cope half.

The following drawings will show you how to make the pattern. It's made of 1/4" plywood. All vertical surfaces on the pattern are given at least a 2 degree draft angle in the direction shown by the arrows. Remember to fillet all inside corners with auto body putty, sand the surface of the pattern smooth and apply at least a couple of coats of gloss varnish or polyurethane.

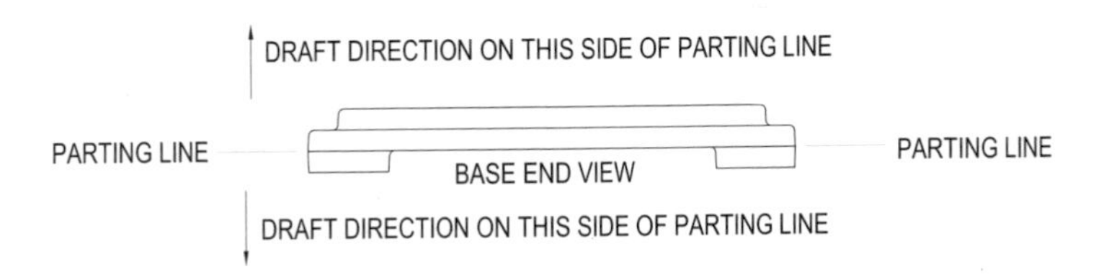

Figure 10 *The base is a split pattern with the legs located in the cope half and the base in the drag half. The end view drawing above shows the location of the parting line. Note the draft direction in opposite directions at the parting line as shown by the arrows.*

Figure 11 *Step 1. Make the outside frame of the mounting base from 1/4" plywood. Note draft direction.*

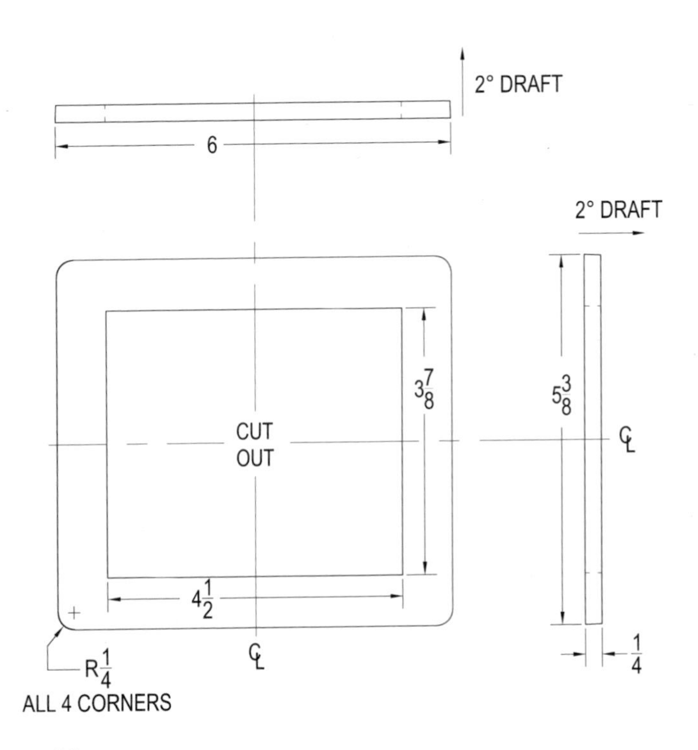

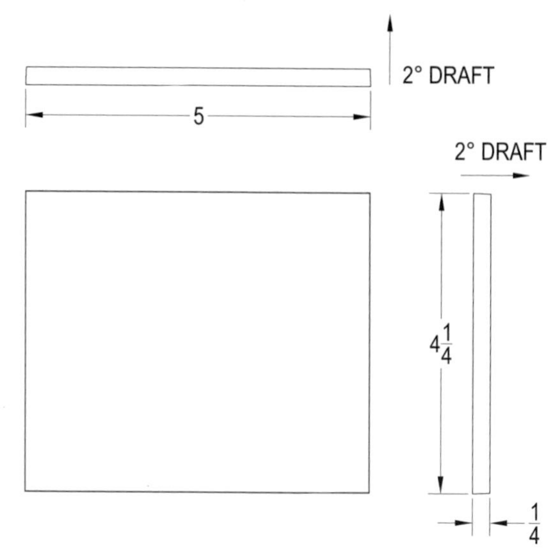

Figure 12 *Step 2. Make the base mounting pad. Made from 1/4" plywood.*

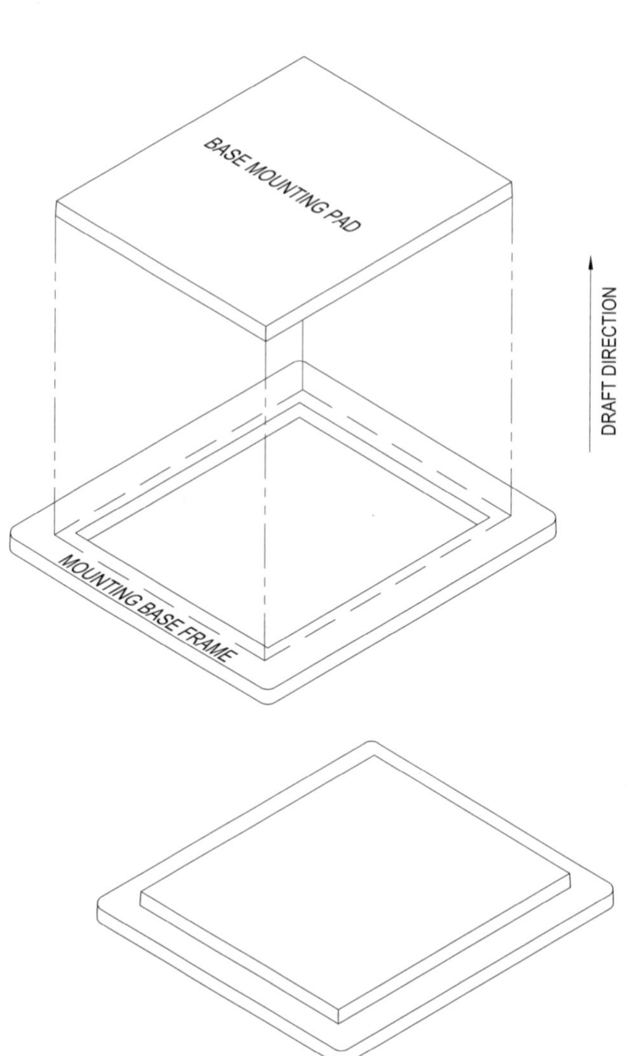

Figure 13 *Step 3. Apply a bead of wood glue around the bottom edge of the base mounting pad. Then position it centered on the mounting base frame. Clamp the two together until the glue sets. Once again, the arrow in the drawing shows the draft direction .*

Figure 14 *The mounting base ready for the legs.*

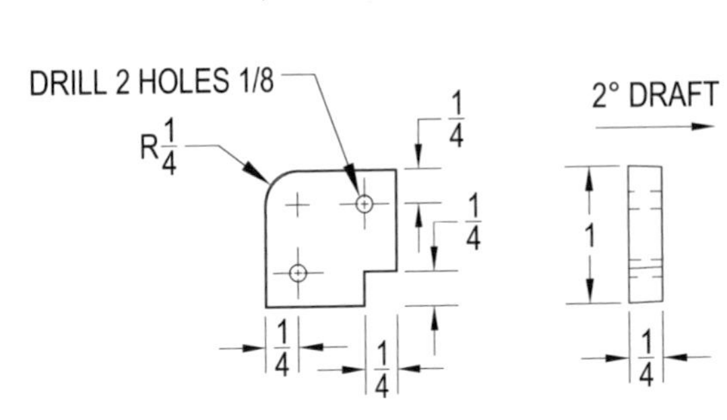

Figure 15 *Step 4. Make 4 legs for the base from 1/4" plywood.*

Figure 16 *Step 5. Since this is a split pattern, the molding procedure will require that the legs be removable from the base. Dowel pins are glued to each leg with corresponding holes drilled in the bottom of the base for positioning the legs. In the drawing at the right we are locating the position of the pin holes for the legs in the bottom of the base.*
To accomplish this, position a leg on a corner of the underside of the base. Make sure the sloping sides of each leg point in the draft direction shown by the arrow in the drawing. Align the outside edges of the leg with the outside corner of base. Use a pencil or punch to transfer the location of the 2 holes from the legs onto the base. Set the leg to one side, then drill corresponding 1/8" holes 3/16" deep into the base. Prepare all 4 corners of the base in the same manner. Number the legs and their position on the base so you can locate them in the proper position later.

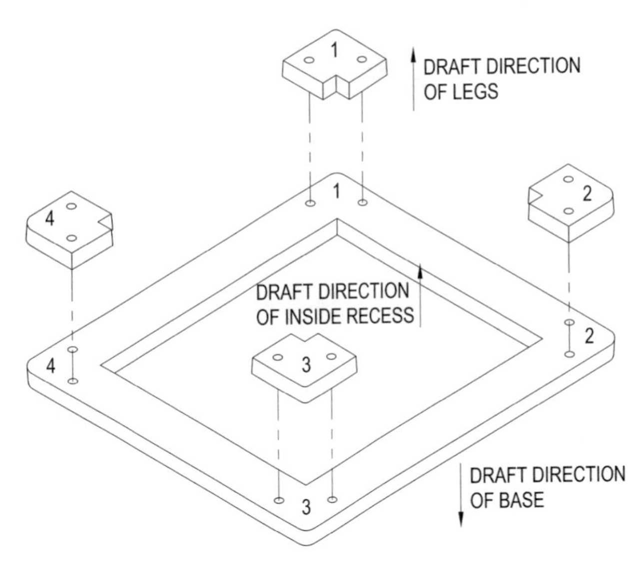

Figure 17 *Step 6. Glue 1/8" diameter wooden dowels into the 1/8" holes of each leg. One end of the dowel to be flush with the surface on the draft end and the other end to extend out 3/16".*

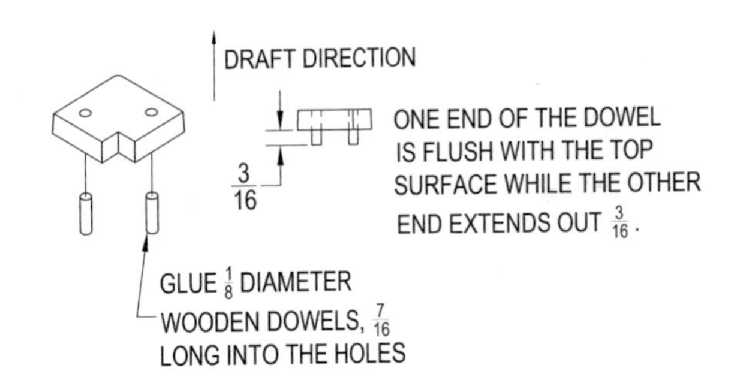

Molding procedure for the base . . .

A step by step molding and casting procedure for the base is shown on the next couple of pages. This will give you a better idea of what the molding procedure is all about. And it will be much the same for the other items cast in this project.

Figure 18 *The base pattern. The legs are beside the base. Notice each one is numbered.*

Step 1. *The base pattern is positioned on the bottom board and centered in the drag. The pattern and bottom board surface are being given a light dusting of parting compound.*

Step 2. *Fill and ram molding sand into the drag.*

Step 3. *Use a straight edge to strike off the surface level with the top edge of the drag.*

Step 4. *Vent the mold..*

Step 5. *Rub in a bottom board.*

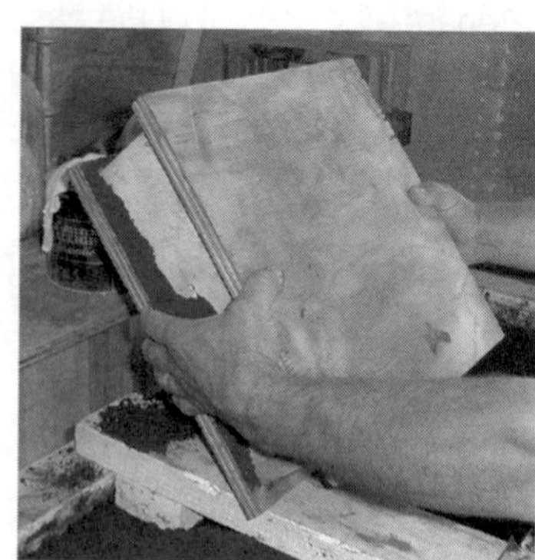

Step 6. *Roll over the drag.*

Step 7. *This is a split pattern with the base located in the drag and the legs for the base located in the cope. Here I am placing the legs on the base pattern*

Step 8. *Set the sprue in position to one side of the pattern.*

Step 9. *Set the cope in place and give the mold and pattern surface a dusting with parting compound.*

Step 10. *Fill and ram molding sand into the cope.*

Step 11. *Strike off the surface of the cope, vent the mold again in 3 or 4 places using vent wire like in step 4. Remove the sprue and procede to step 12.*

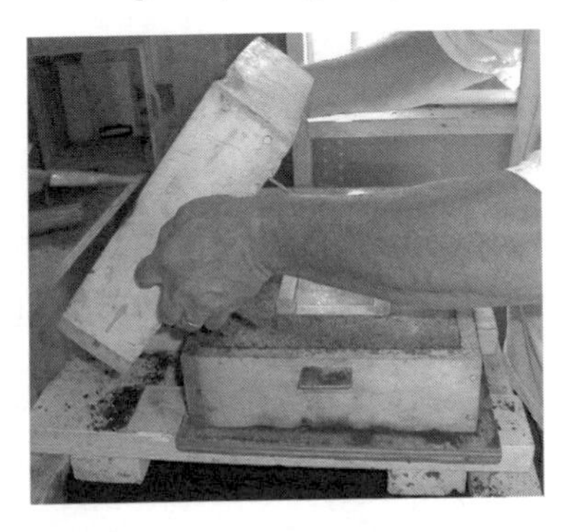

Step 12. *Lift the cope from the drag and set it to one side.*

Step 13. *Using a wet soft bristle brush to swab the edges of the mold cavity. This strengthens the edges making them less likely to crumble when the pattern is removed.*

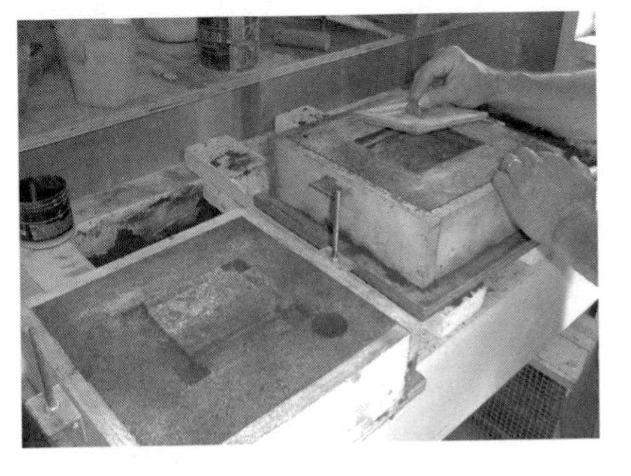

Step 14. *The pattern half being removed from the mold.*

Step 15. *Cutting the gate.*

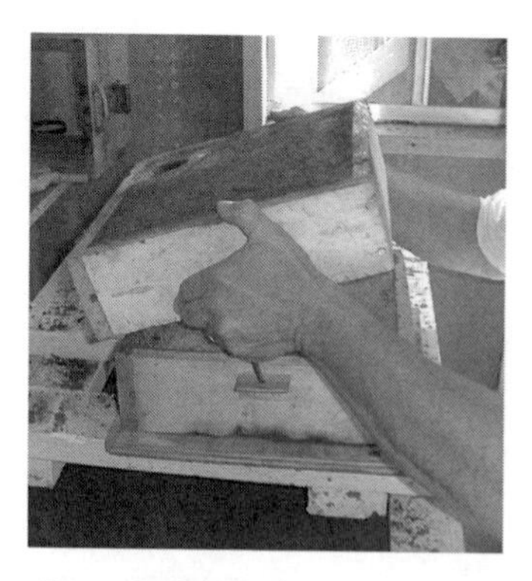

Step 16. *Reassemble the mold and your ready to pour.*

Step 17. *Pouring the mold.*

Step 18. *The finished casting with sprue still attached. The next step will be to cut away the sprue and clean up the casting by removing the rough edges with a file.*

The side panel pattern . . .

This is a single pattern made from 1/4" plywood and 3/8" x 3/8" pine. As in all patterns, pay attention to the draft angle when making the pattern. Fillet all inside corners then sand the pattern smooth and finish with a couple of coats of polyurethane or gloss paint.

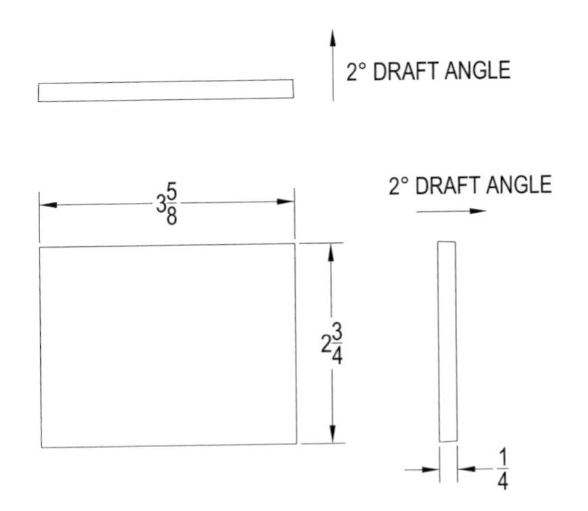

Figure 19 Step 1. Make the side panel pattern from 1/4" plywood.

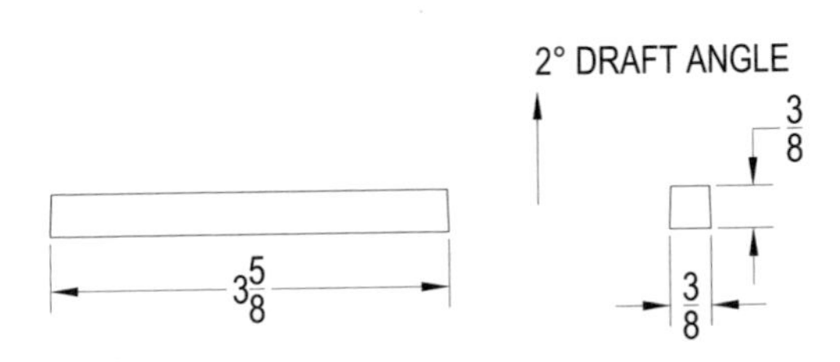

Figure 20 Step 2. Make the 2 rails for the side panel pattern from 3/8" x 3/8" pine.

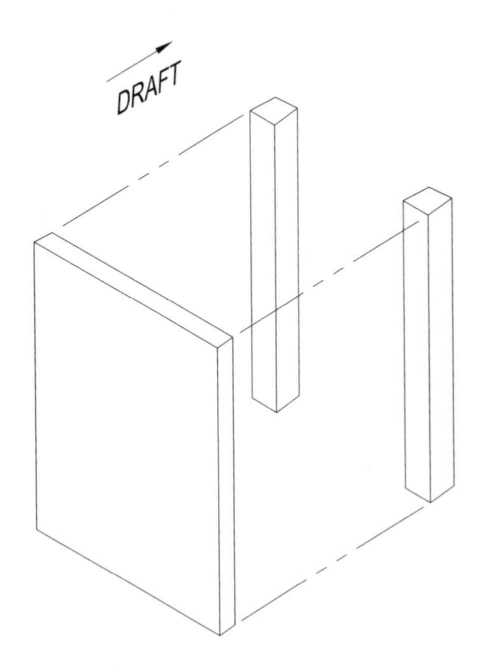

Figure 21 Step 3. Apply a bead of glue to the back side of each rail . Then align the outside edge of each rail with the outside edge of the panel. Clamp the rails in position on the panel and allow the glue to dry. As always, pay close attention to the direction of the draft angle.

Side panel molding procedure . . .

Molding procedure for side panel is similar to that of the base except this is a single pattern. Molding procedure as follows:

Place base of pattern on the molding board. Dust with parting compound, then position the drag on the bottom board so the pattern is centered in the drag. Ram the drag full of molding sand, strike off, vent the mold then rub in a bottom board. Roll over the drag. Press a sprue to one side of the pattern. Dust the mold surface and pattern with parting compound, then set the cope in place over the drag. Ram the cope full of molding sand then strike off. Remove the sprue. Lift the cope from the drag and set it to one side. Remove pattern, cut the gate, clean up mold cavity, reclose the mold and pour.

Figure 22 *The completed side panel pattern.*

Front & rear panels . . .

The size and shape of the front and rear panel blanks shown in figure 23 are the same. Make both from 1/4" thick plywood. The final preparation of the panels will be different though.

Prepare the front panel first. Looking at figure 24, locate and drill the three 1/8" holes. Then locate and make the cut out as shown.

Next make the water jacket mounting pad shown in figure 25 from 1/4" plywood.

Then make 2 pivot bosses as shown in figure 26 from 1" diameter wooden dowel.

Then make two crank shaft bosses as shown in figure 27 from 1-1/4" wooden dowel. One of the crank shaft bosses will be used on the front panel and the other on the rear panel. The draft angle on the bosses can be achieved on the lathe. Each boss is drilled 1/8" through the center and then an 1/8" dowel is glued in as shown in figure 28. The dowel end protruding out the bottom comes in handy when locating the bosses on the panels.

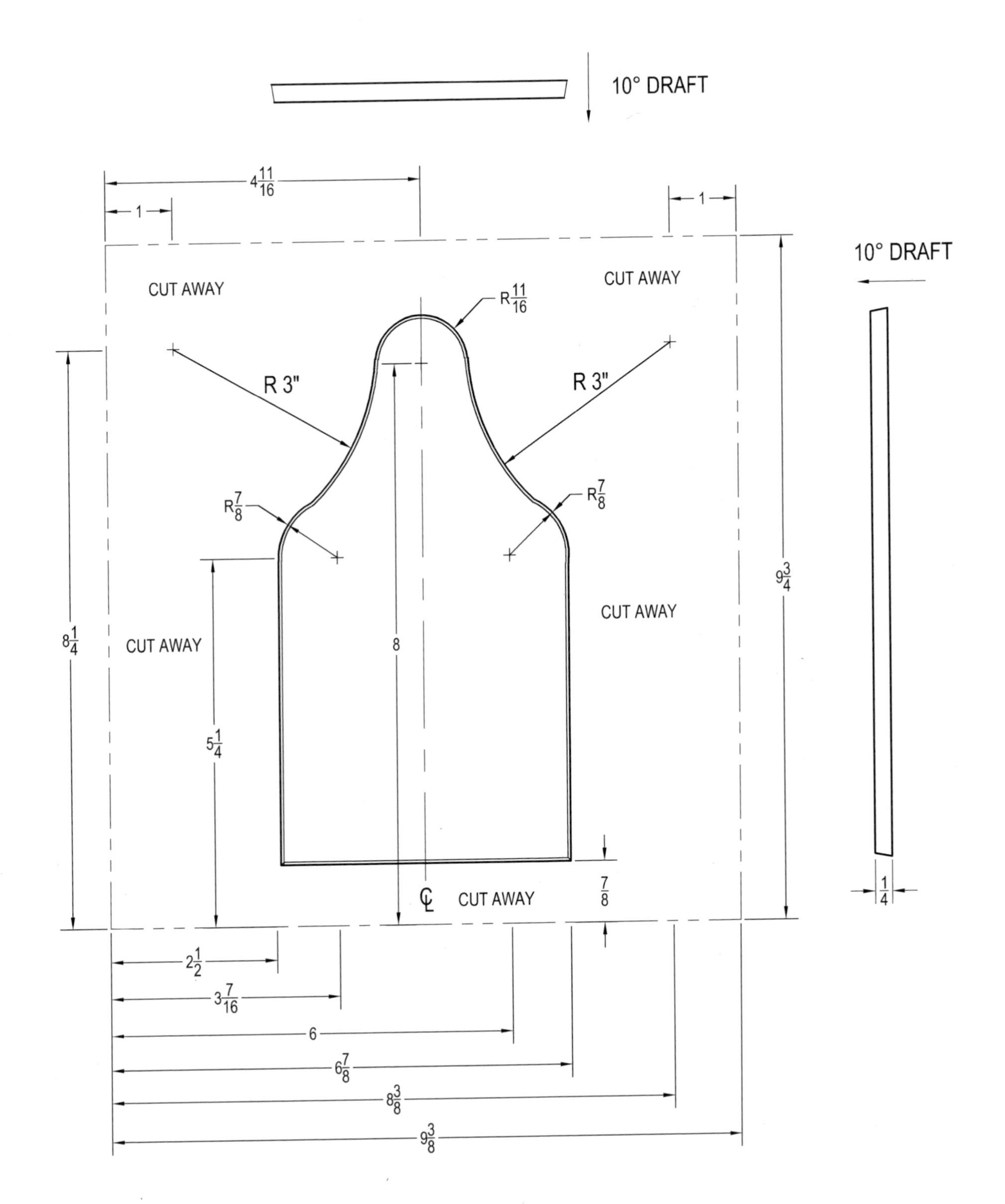

Figure 23 *The front & rear panel blank. Make 2. Each one cut from a piece of 1/4" plywood measuring 9-3/8" x 9-3/4".*

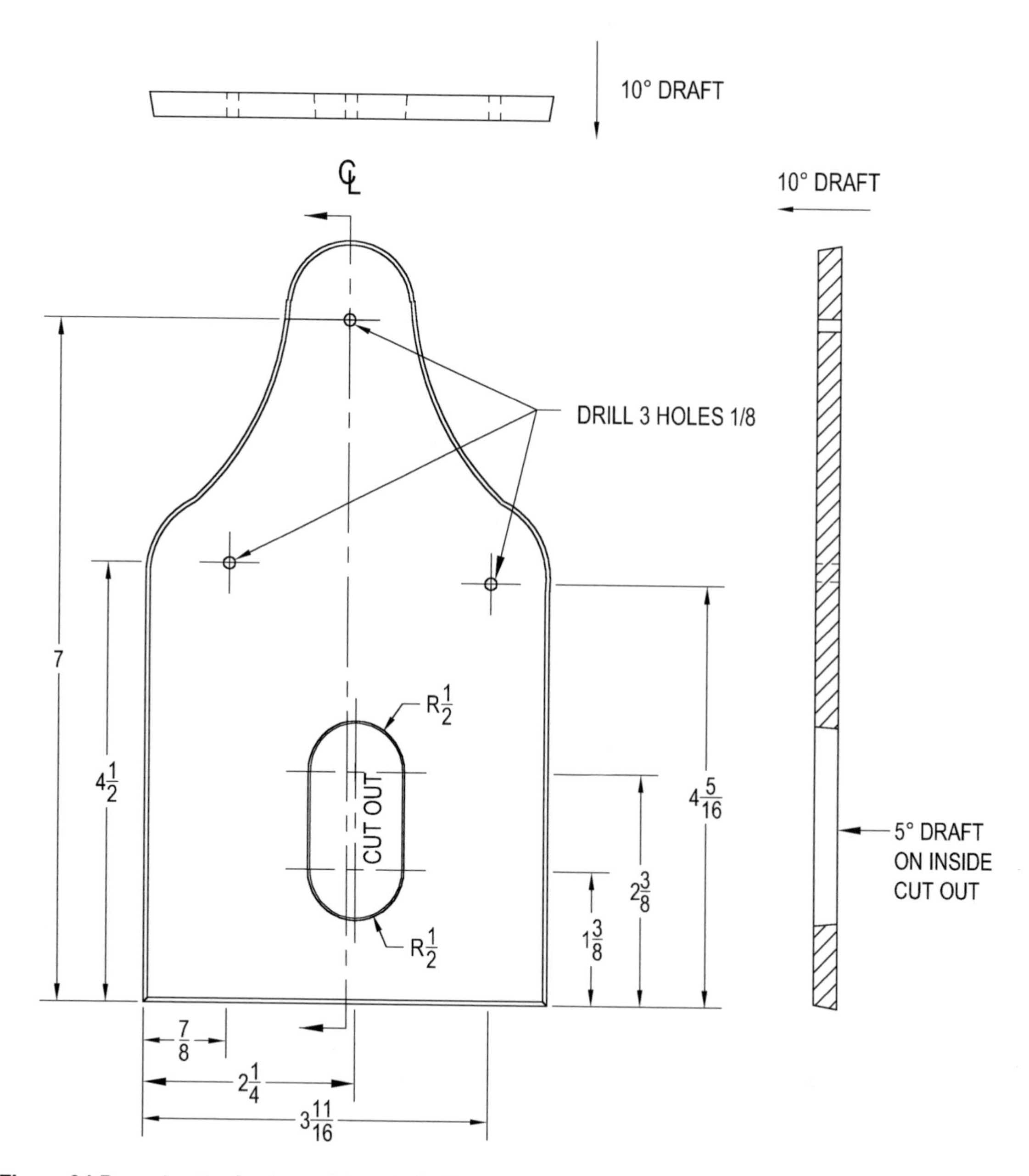

Figure 24 *Preparing the front panel. Locate & drill the three 3/16" holes and make the cut out..*

Coat the ends of the bosses with wood glue then mount them in position on the front panel as shown in figure 29. The 1/8" wooden dowel end of each boss is inserted in the 1/8" hole in the front panel.

Glue the mounting pad to the front panel next. Align the mounting pad cutout with the cutout in the front panel, then clamp in position until the glue sets. You will notice that the cylinder mounting pad is located slightly off center. This is because in final assembly, the cylinder will likely be located slightly off center.

We wanted to add a raised edge to our casting to add an appearance of authenticity and style. We accomplished this by gluing 14 gauge single strand insulated

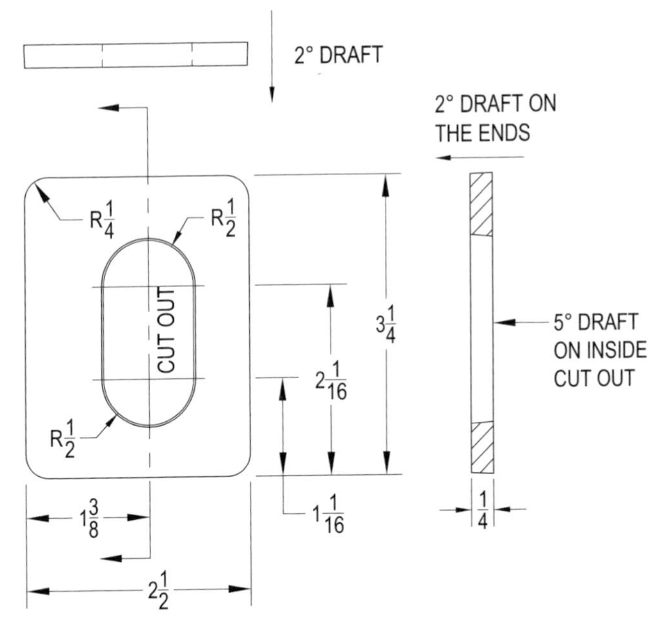

Figure 25 Water jacket mounting pad. Make one from a piece of 2-1/2" x 3-1/4" x 1/4" plywood.

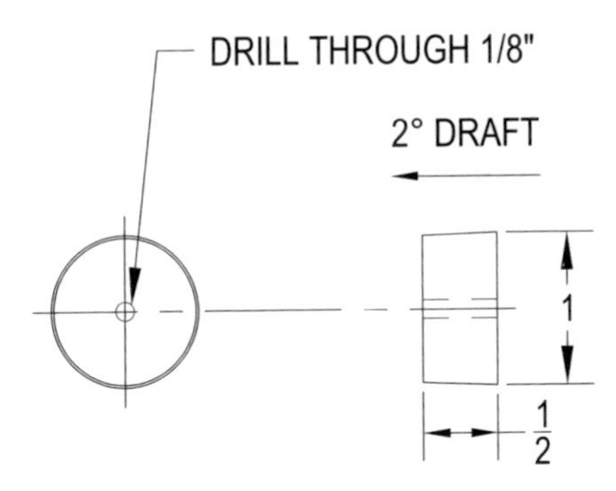

Figure 26 Pivot boss. Make two for the front panel from 1" wooden dowel.

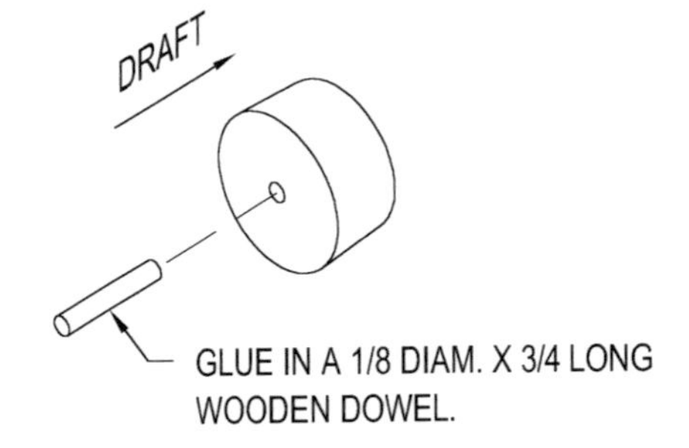

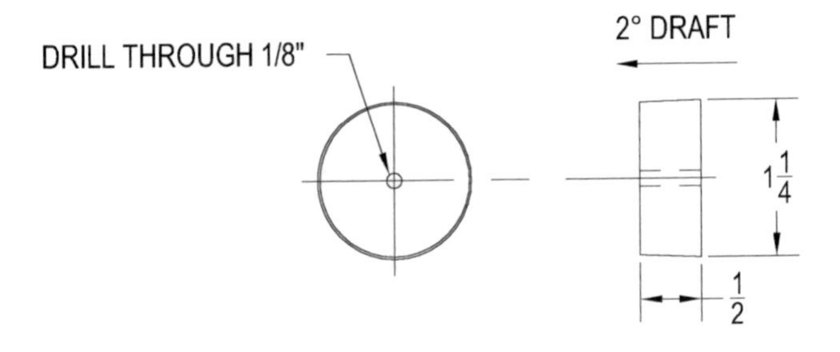

Figure 27 Crank boss. Make two, one for the front panel and one for the rear panel from 1-1/4" wooden dowel.

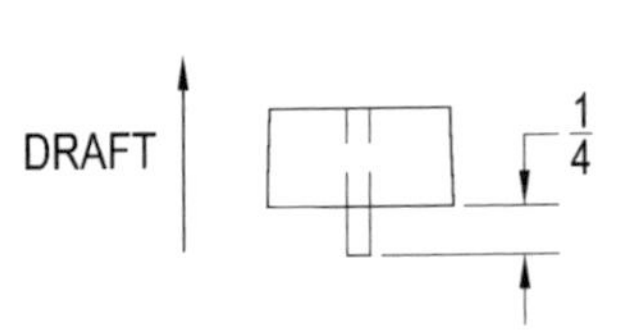

Figure 28 Alignment pins are made from 1/8" dowel and glued in each boss as shown.

wire along the outside edge of the pattern as shown in figure 30. A good all purpose all hold glue works well for this. You can hold the wire in position with small clamps until the glue sets.

After the pattern is assembled, fillet the inside corners, sand smooth and finish with a couple of coats of polyurethane.

Prepare the rear panel next. Begin by gluing the crank boss in position as shown in figure 32. Next glue the 14 gauge wire along the edge as shown in figure 33. Finally fillet all inside corners and finish with a couple of coats of polyurethane.

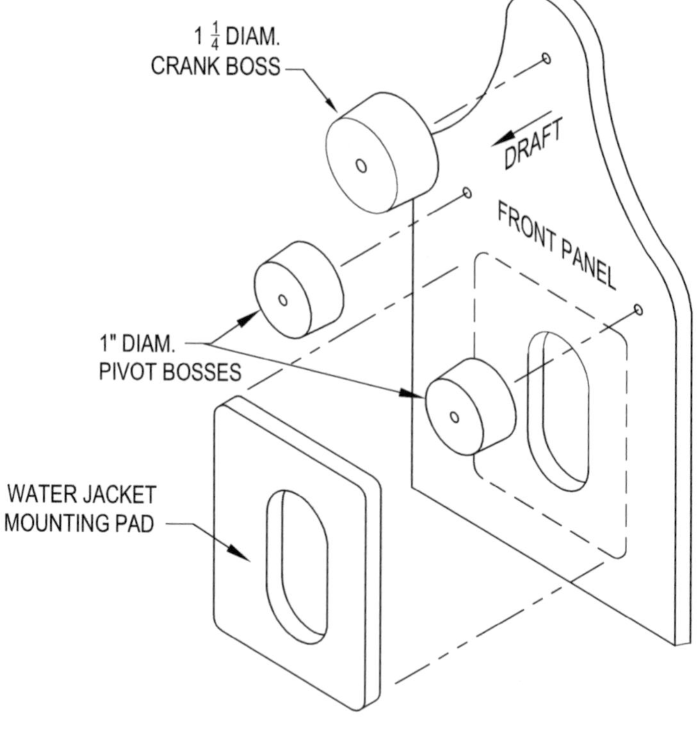

Figure 29 *Mount the crank boss, pivot bosses and water jacket mounting pad to the front panel pattern.*

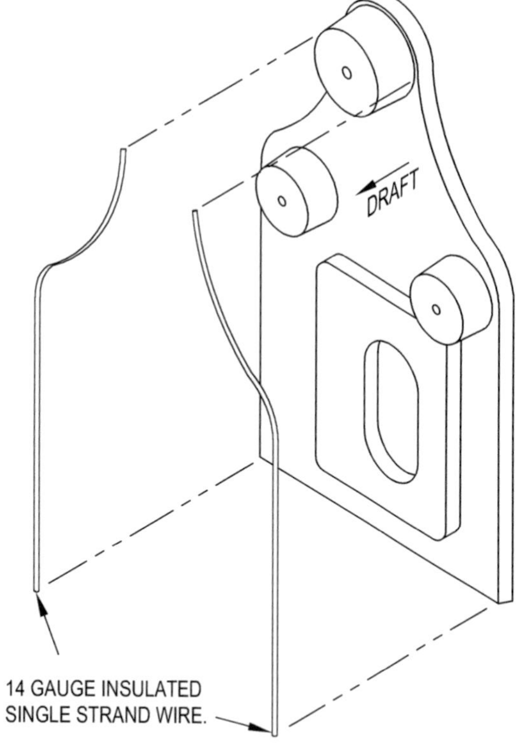

Figure 30 *Gluing 14 gauge wire along the outer edge of the front panel pattern. I think the raised edge this produces on the final castings really adds to style and authenticity of the look.*

Figure 31 *The completed front panel pattern.*

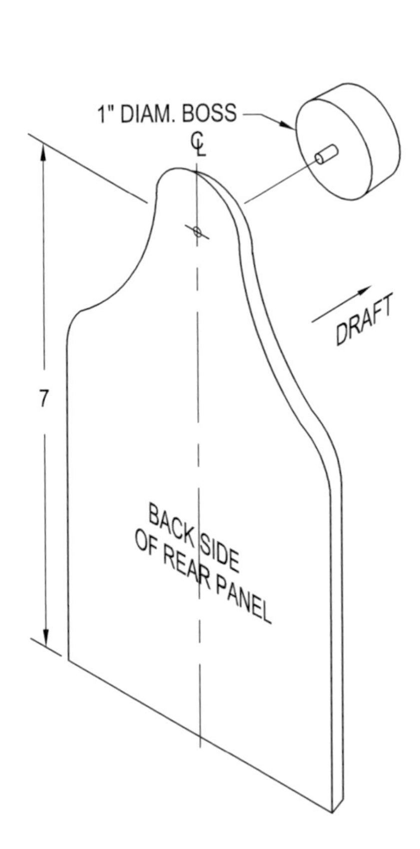

Figure 32 *Mount the crank boss to the rear panel pattern.*

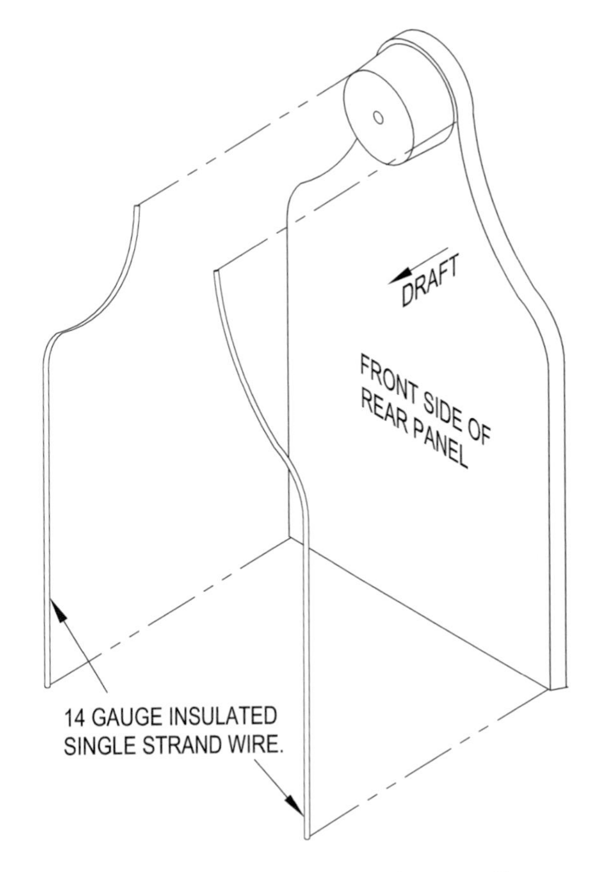

Figure 33 *Gluing 14 gauge wire along the outer edge of the rear panel pattern.*

Figure 34 *The completed rear panel pattern.*

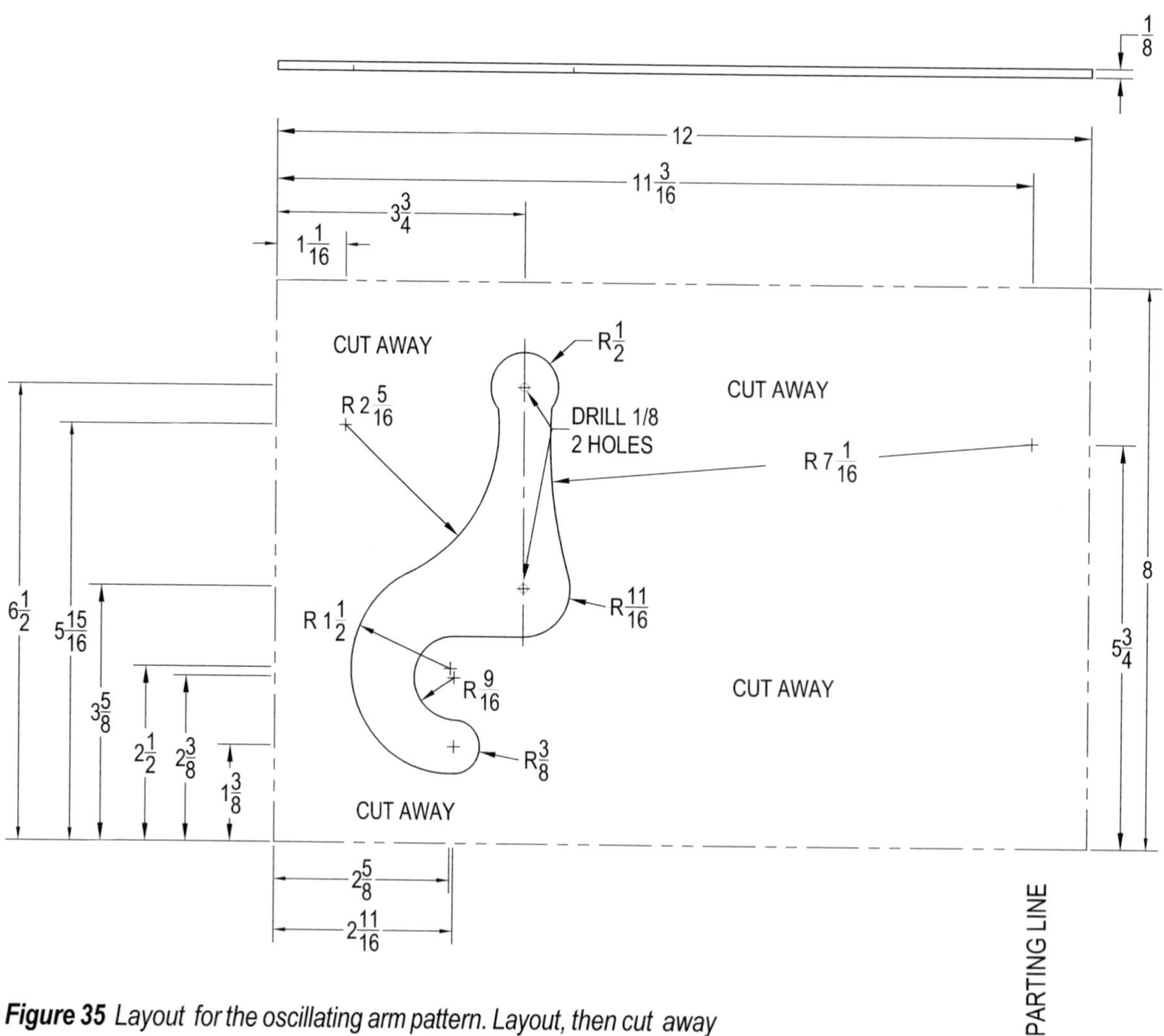

Figure 35 Layout for the oscillating arm pattern. Layout, then cut away the pattern from a 12" x 8" piece of 1/8" masonary board. Mark the word pattern on the cut out. Use the newly created pattern to create 4 more identical shapes from 1/8" masonary board.

Oscillating arm patterns . . .

A pair of oscillating arm patterns are needed. One left, one right. The patterns will be split patterns which of course means the pattern will be in 2 parts with one half of the pattern located in the drag and the other half in the cope so you will need 4 identical pieces of the arm shape detailed in figure 35. Each piece will be a pattern half and is cut from 1/8" thick masonry board. Rather than going through the process of laying out each

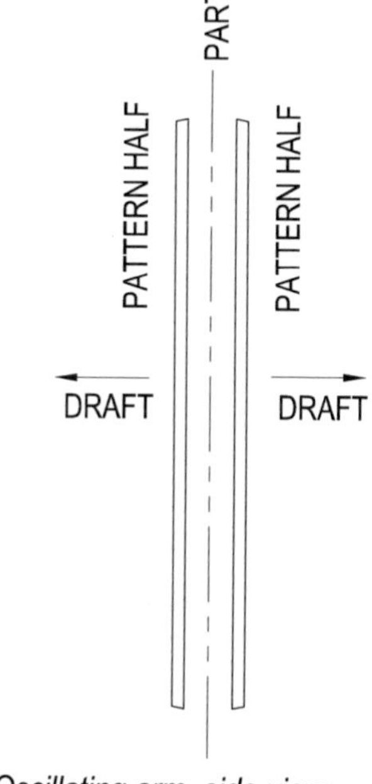

Figure 36 Oscillating arm side view.

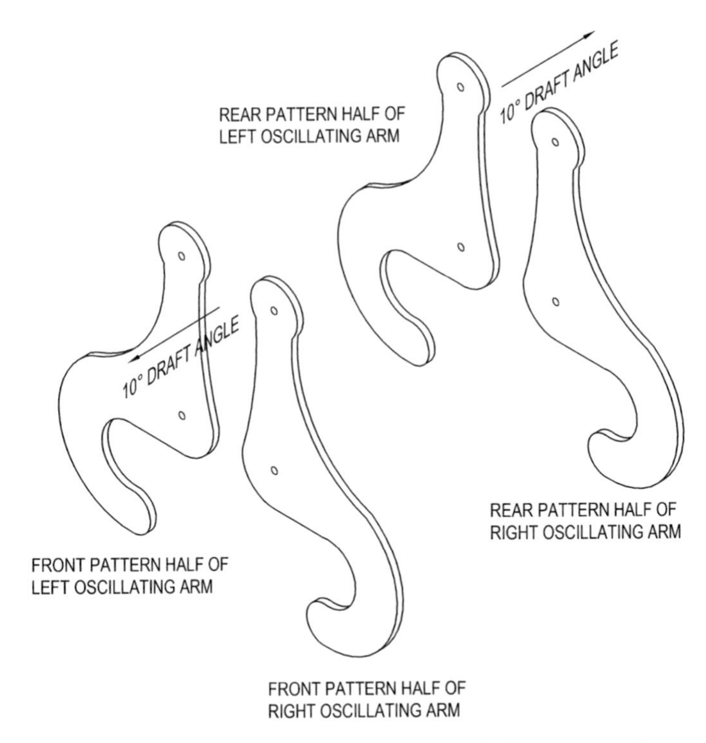

Figure 37 *How the oscillating arm pattern halves go together.*

pattern half, the simplest way would be to make a pattern from the dimensions in figure 35. Mark it so you know it's the pattern. Then use the pattern to mark and cut the 4 pieces needed.

After the pieces have been cut out, add a draft angle to each piece. A belt sander works good for this. Figure 37 shows how the pieces go together to form the complete pattern while figure 36 shows the opposing draft angles of the pattern halves. As before, the arrows denote the draft angle.

The next order of business is to make and attach the bosses and alignment pegs on the pattern halves.

You'll need to make 3 sets of bosses as shown in the figures. I turned the draft angle on each using the lathe.

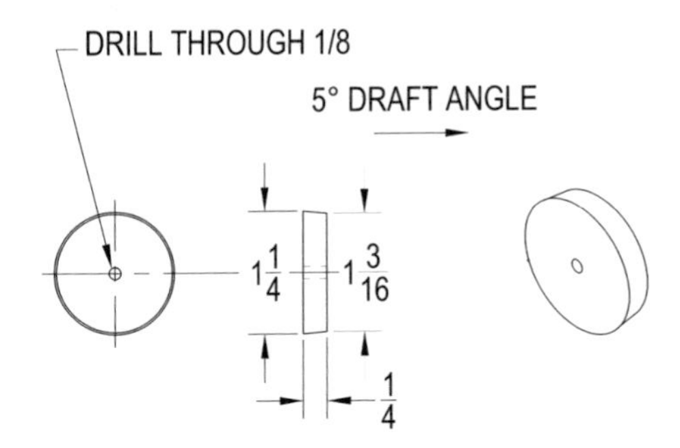

Figure 38 *Oscillating arm pivot boss for front pattern halves. Make 2 from 1-1/4" diameter wooden dowel.*

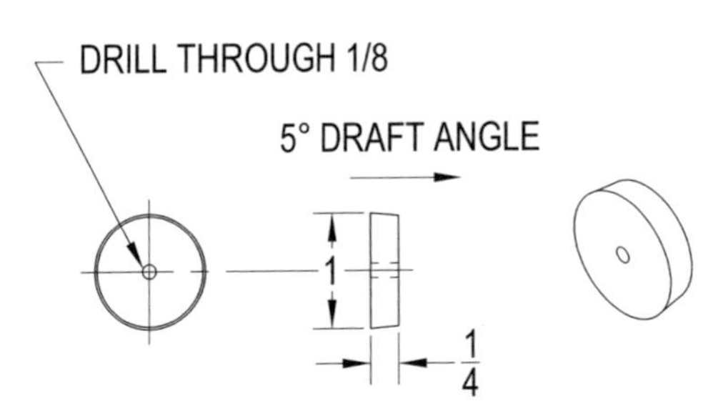

Figure 40 *Oscillating arm crank boss for front pattern halves. Make 2 from 1" diameter wooden dowel.*

Figure 39 *Oscillating arm pivot boss for rear pattern halves. Make 2 from 1-1/4" diameter wooden dowel.*

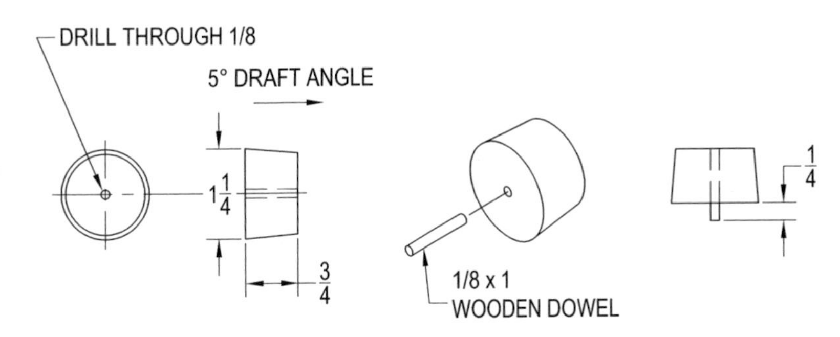

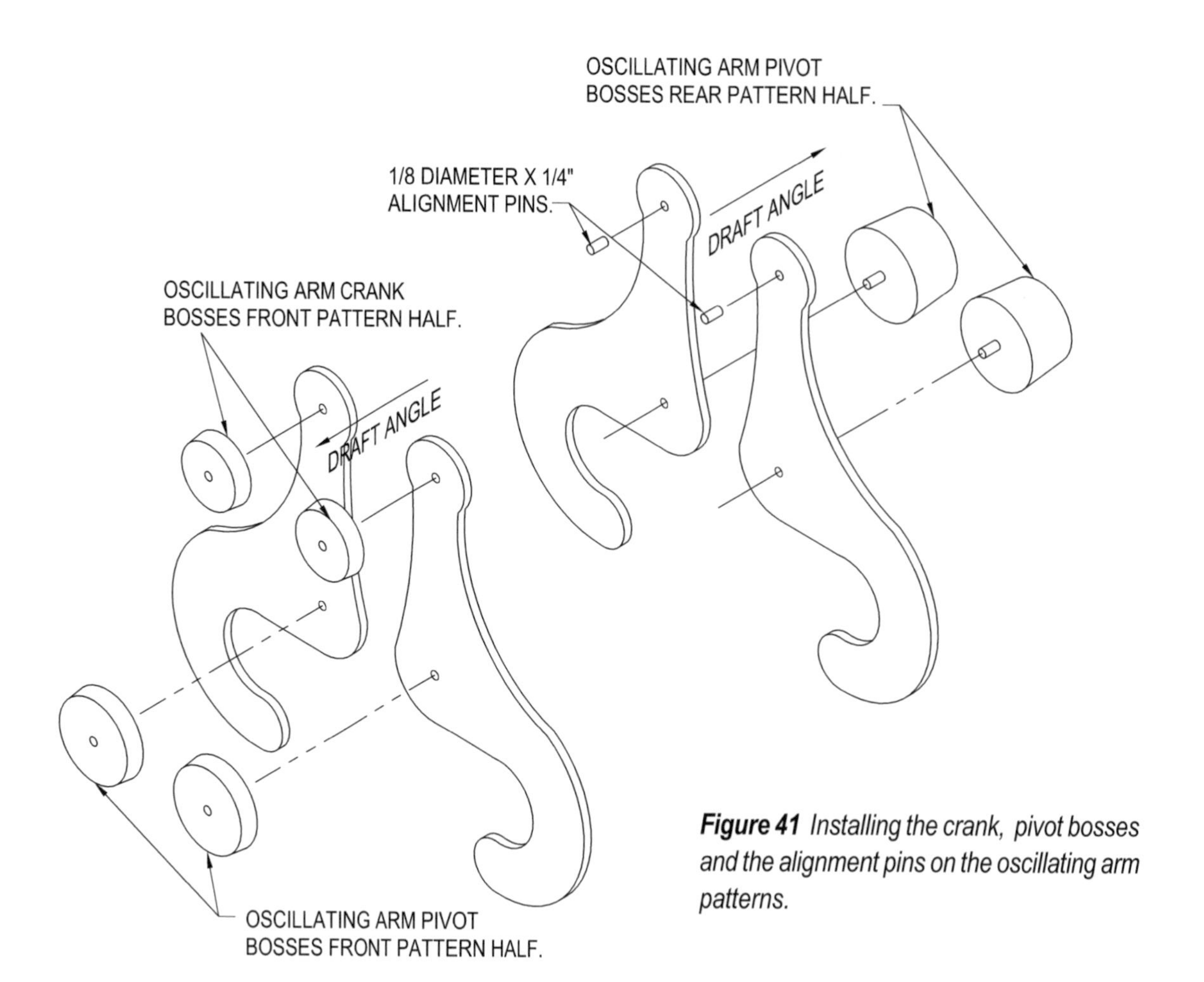

Figure 41 *Installing the crank, pivot bosses and the alignment pins on the oscillating arm patterns.*

The bosses are installed on the pattern halves as shown in figure 41. Start by applying a small amount of wood glue to the back side of the front pattern bosses. Align the 1/8" holes in the bosses with the 1/8" holes in the front pattern half then clamp the bosses in position until the glue hardens. The 1/8" holes in the front pattern half are the receiver holes for the alignment pins in the rear pattern half.

Apply a small amount of wood glue to the pivot bosses for the rear pattern half. Insert the 1/8" dowel located on the end of each boss into the 1/8" holes in the rear pattern half. The dowel should protrude at least 1/8" out the other side of the pattern half.

Next make the two 3/4" long alignment pins from 1/8" wooden dowel rod. Apply a bit of wood glue to the end of the pins and

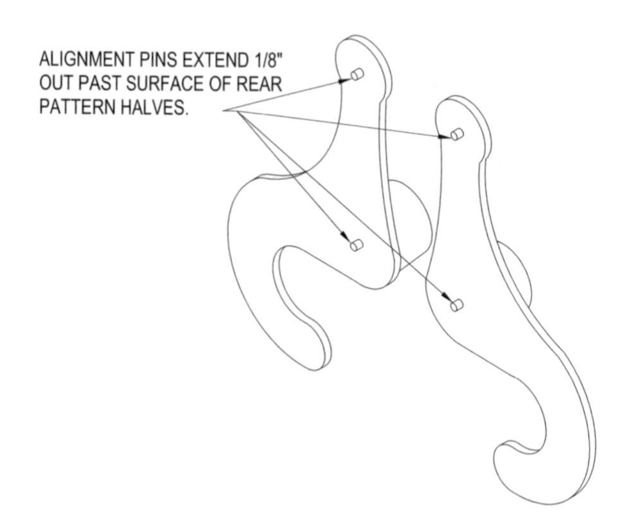

Figure 42

insert them into the rear pattern half as shown in the figure. The pins should protrude out the back side of the rear pattern halves about 1/8" as shown in figure 42.

Here again, you might want to add a bit of character to your castings by adding the raised bead around the edges. 14 gauge single strand insulated wire is bent to shape and glued around the outside edges of the front pattern halves as shown in figure 43.

The next steps are to fillet all inside corners with auto body putty. On the front pattern halves you will have to fillet around the wire to smooth the transition. Then sand all surfaces smooth and coat with polyurethane.

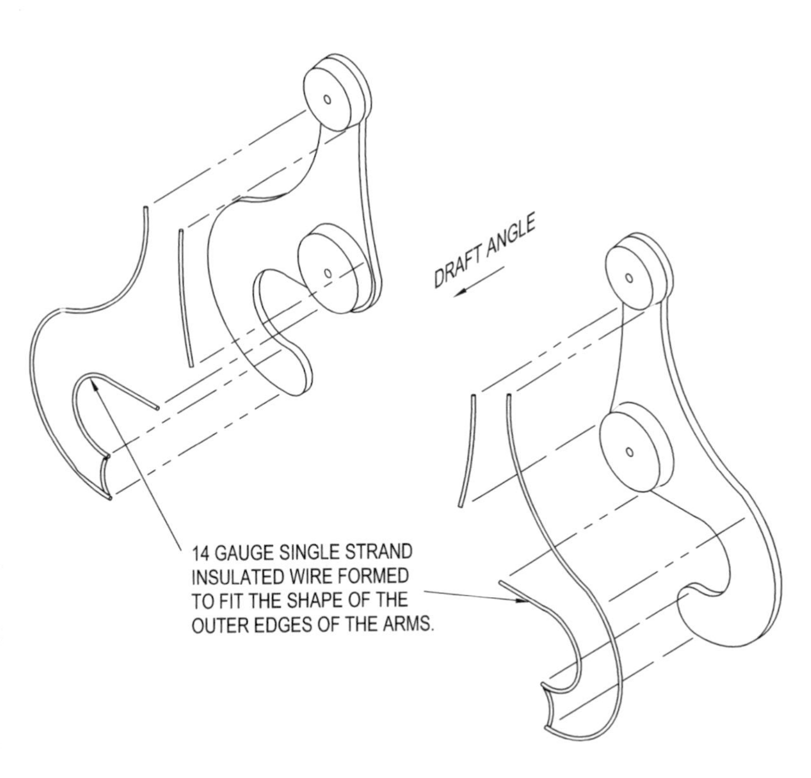

Figure 43 *Applying the raised bead to the oscillating arm front pattern halves.*

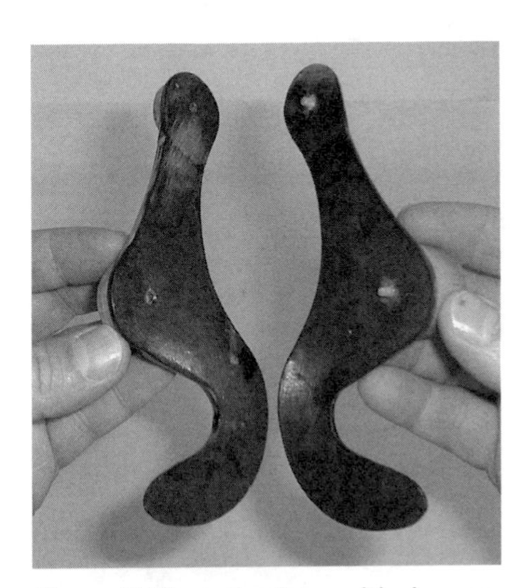

Figure 44 *The split pattern as it looks separated. The front pattern half is the one on the left with the reciever holes for the alignment pins. The rear pattern half is the one on the right with the alignment pins.*

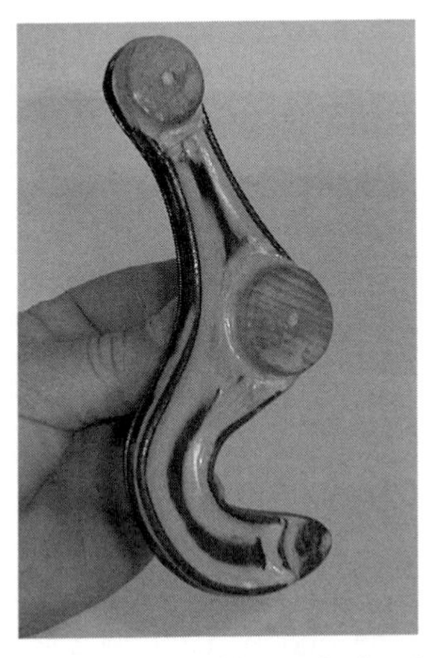

Figure 45 *Looking at the front of the oscillating arm pattern. Notice how all the inside corners have been filleted. Also, the 1/8" holes in each boss have been filled.*

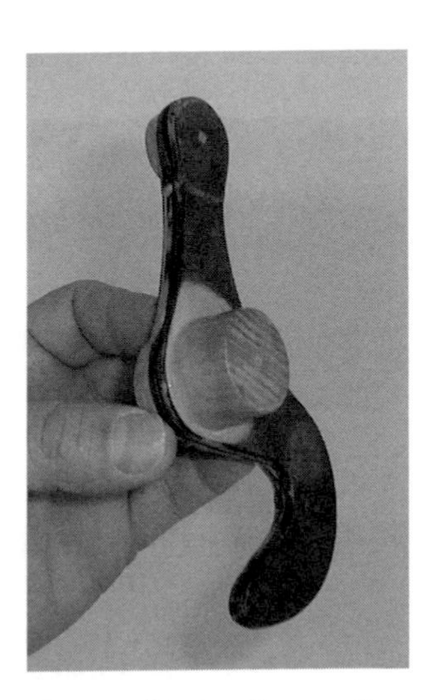

Figure 46 *Looking at the rear of the oscillating arm pattern. Here, the inside corner around the pivot boss has been filleted.*

Piston rod pattern . . .

The piston rod pattern is another split pattern. The pattern is designed so that 2 piston rods could be machined from the casting created from it.

Make the pattern halves as shown in figure 47 from pine, fir or other suitable material. When completed, prepare the pattern halves with alignment pins as shown in figure 48. Finally, sand the pattern smooth and coat with polyurethane.

Casting procedure for this pattern is the same as with previous split patterns.

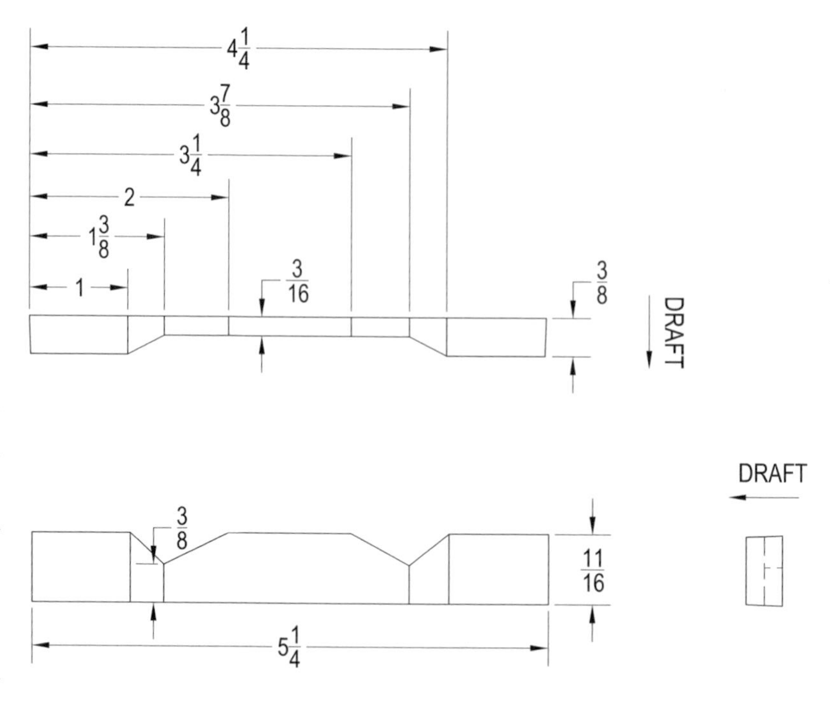

Figure 47 *Piston rod pattern detail. Make 2 halves.*

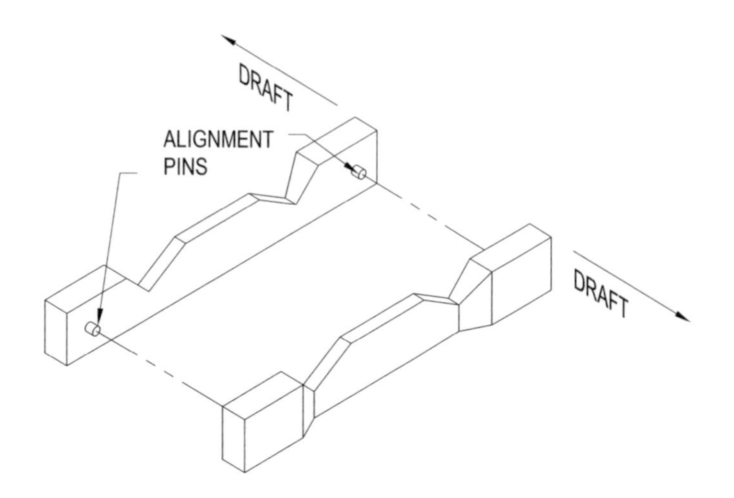

Figure 48 *Location of alignment pegs in pattern halves.*

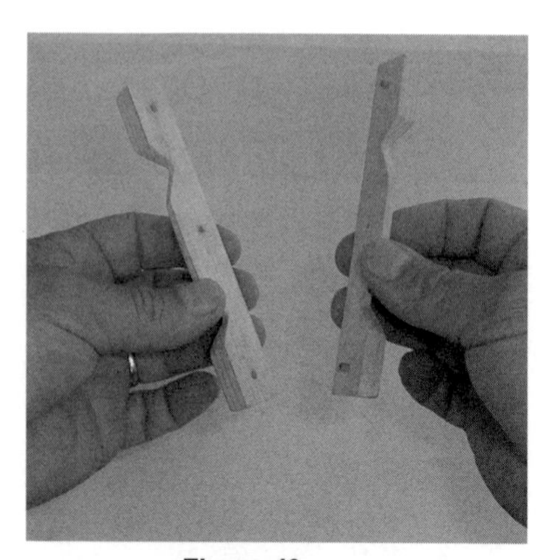

Figure 49

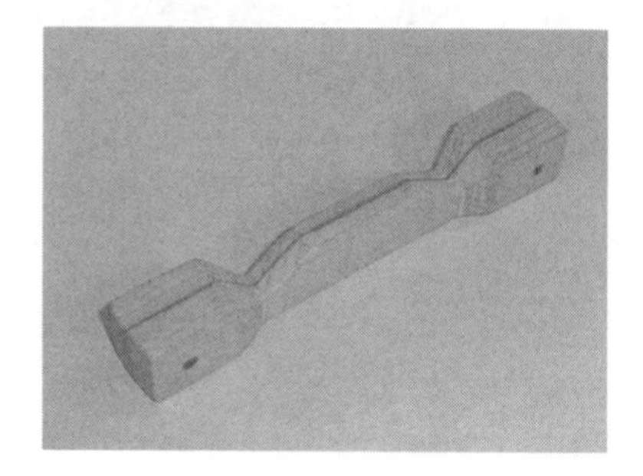

Figure 50

The flywheel pattern . . .

The flywheel pattern design we're using is much the same as one we have used on several of our previous engine projects. It's a two piece split pattern and it's easier to make than what you might think at first glance. The two halves are simple flat back patterns split at the natural parting. A pair of pegs align the patterns in perfect register.

The original pattern was made from 5/8" thick, C grade, exterior fir plywood and has held up for several years now and has produced dozens of molds and is still in excellent condition. Fir plywood is not the best material for making patterns, but by filling the imperfections with auto body putty, doing a little sanding and then finishing with a couple of coats of varnish you can create an acceptable pattern.

The pattern layout is shown in figure 51. The method used to do the layout is a simple sequence done with a compass. Small washers can be used to mark the fillets. The pattern consists of two 9" diameter plywood disks. It is best to mark the spoke pattern on one of the halves and then clamp the two together to cut out the spoke openings. By doing it in this manner you will have only one layout and are assured a matched pair.

First, divide the 9" circle into 6 equal parts. This can be done with a protractor or a 60 degree triangle. Use a compass to mark the 7-3/4" circle representing the inside radius of the outer rim. Next mark the 6-3/4" circle representing the inside rim. Referring to step 2 in figure 52, set the compass at 3-3/8" which is the radius of the inner circle. Pivot the compass at point A and draw arc AB. Continue around the circle drawing an arc at each division line.

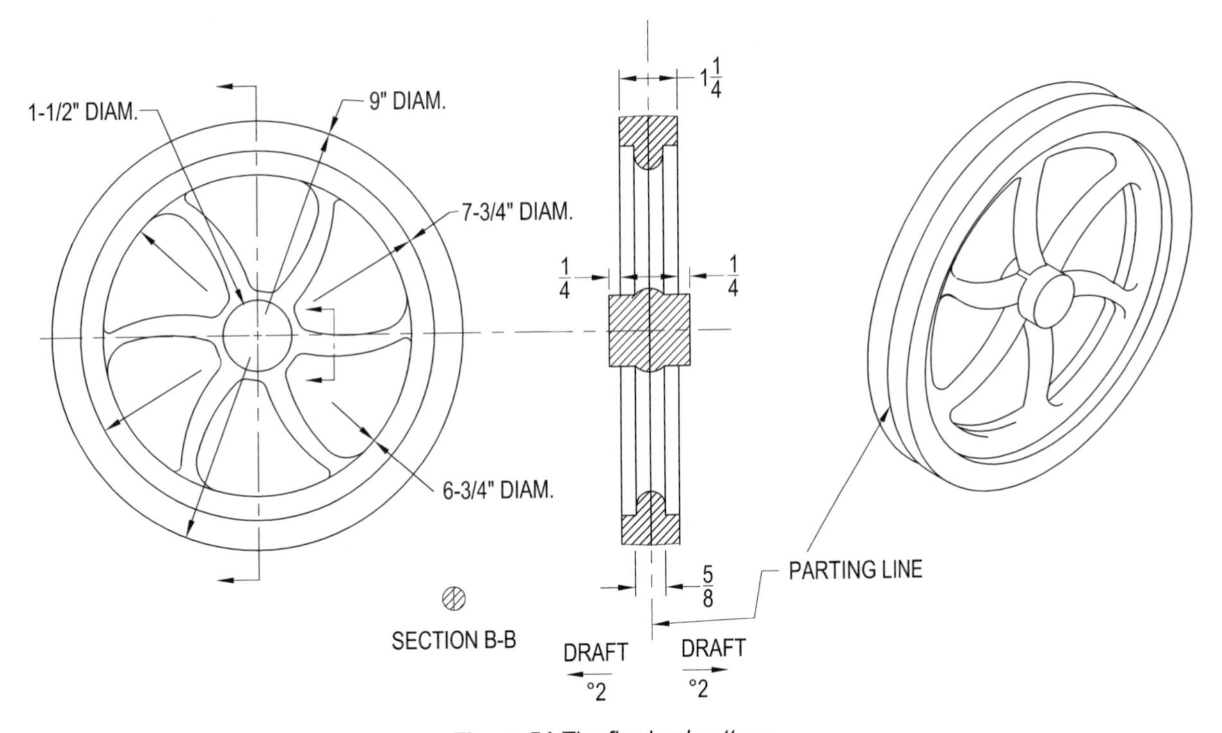

Figure 51 *The flywheel pattern*

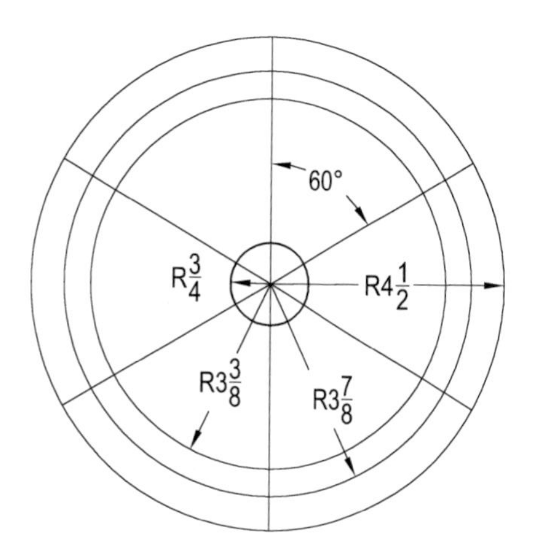

STEP 1. START WITH A 9" DIAMETER WOODEN DISK CUT FROM 5/8" THICK PLYWOOD. DIVIDE THE DISK INTO 6 EQUAL PARTS. DRAW THE 3 CIRCLES ON THE DISK WITH A COMPASS.

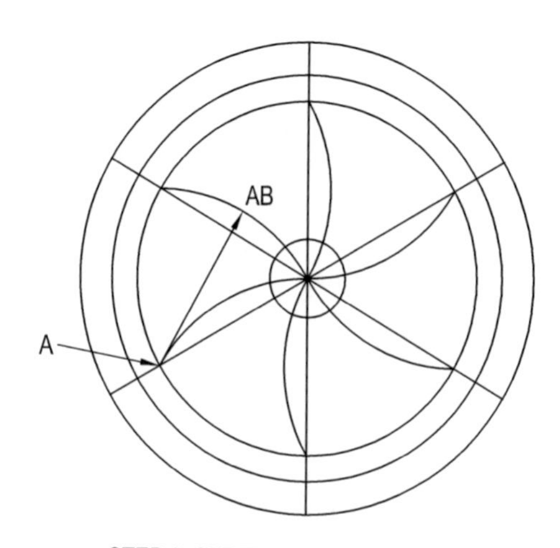

STEP 2. SET THE COMPASS AT 3-3/8" WHICH IS THE RADIUS OF THE INNER CIRCLE. PIVOT THE COMPASS AT POINT "A" AND DRAW ARC "AB". USE THE SAME METHOD TO DRAW AN ARC AT EACH OF THE OTHER DIVISION LINES.

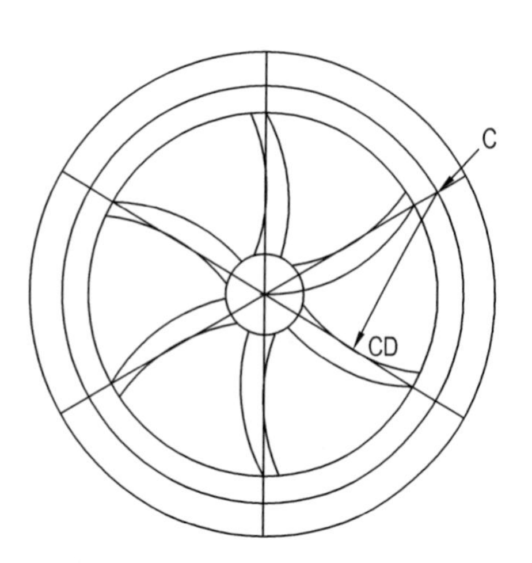

STEP 3. WITH THE COMPASS STILL SET FOR A 3-3/8" RADIUS, PIVOT FROM POINT "C" AND DRAW ARC "CD". CONTINUE AROUND THE CIRCLE DRAWING ARCS AT EACH DIVISION LINE.

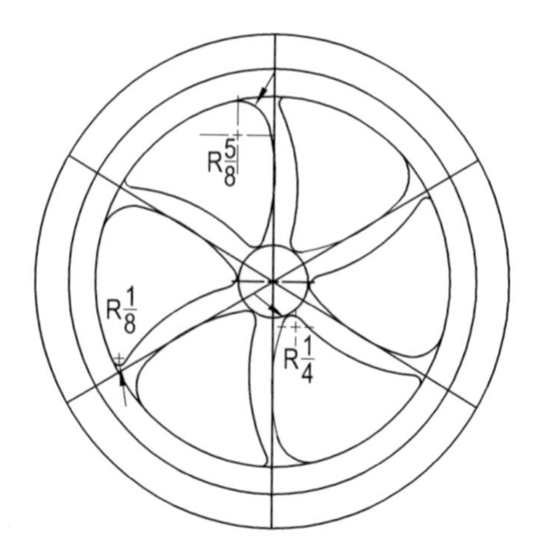

STEP 4. DRAW ARCS TANGENT WITH THE SPOKES AND RIM, AND WITH THE SPOKES AND HUB.

Figure 52 *Layout for flywheel pattern. This is a split pattern so you will be making two halves.*

Referring to step 3 in figure 52, and with the compass still set a 3-3/8", pivot from point C and draw arc CD. Continue around the circle drawing arcs at each division line. Finally draw the fillets tangent with the spokes and rim and spokes and hub. As mentioned earlier the fillets are not critical and can be drawn with washers or other round objects close to the sizes indicated.

With the layout complete, clamp both pattern halves together. Mark them in such a way that they can be reassembled in the exact po-

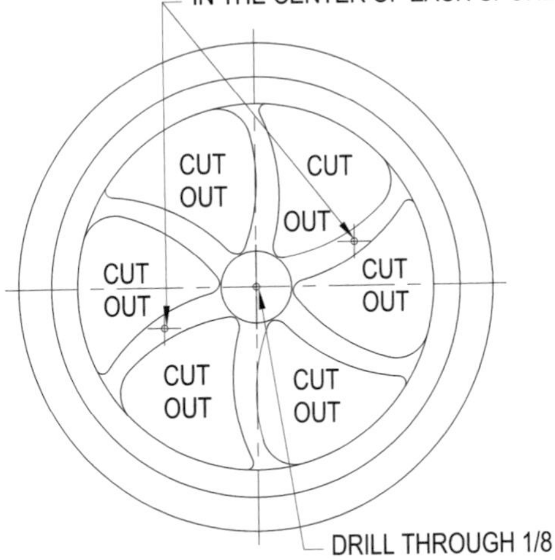

Figure 53 *Clamp the 2 pattern halves together. Drill starting holes then cut out the sections between the spokes. Drill the 1/8" hole through the center hub of both pattern halves and drill the other two 1/8" holes in the spokes for the alignment pegs.*

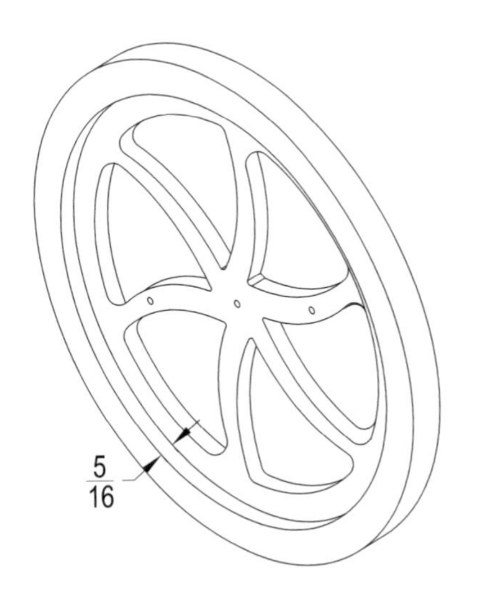

Figure 54 *Mount each pattern half on a lathe face plate to turn the inside rim.*

sition later. Drill starting holes and cut the spoke openings with a jig saw.

Mount either pattern half on a lathe face plate and turn the hub and rim detail.

Center the blank carefully in relation to the original layout center so the pattern will be truly symmetrical. Remember to draft the pattern at each diameter. Make the opposing second half in the same manner.

Now round the spokes and the inside rim as shown in section BB in figure 51. This can be done by hand, or if you have a hand grinder that can be fit with a small sanding drum you can use it to speed the process up a little.

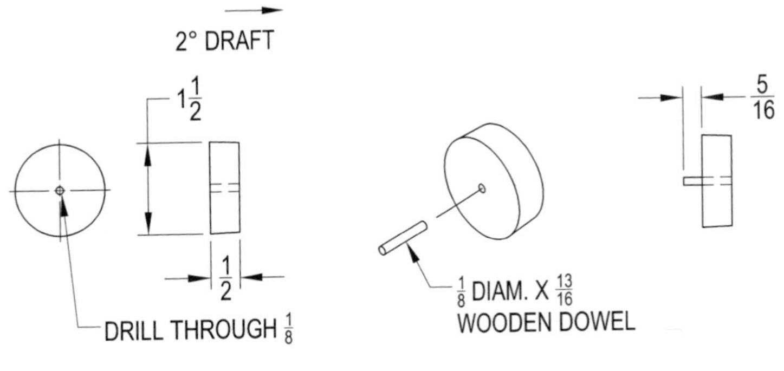

Figure 55 *Flywheel hub. Make 2 from 1-1/2" diameter wooden dowel .*

Finally clamp the halves together in their original position and drill through for the alignment pegs. The alignment pegs can be made from a nail or piece of 1/8" round rod. Insert the pegs in the holes of one pattern half and secure them with epoxy. The location of the alignment pegs is not critical. See figure 56 for an approximate position.

Make the flywheel hubs from 1-1/2" wooden dowel as shown in figure 55. Glue a hub to the center of the outer facing side of each pattern half as shown in figure 56.

Finally, fillet inside corners, Fill any flaws, sand the pattern smooth and seal with two or more coats of varnish.

Lay the pattern half without pegs face down on bottom board. Sprinkle with parting dust and ram the drag up over it. Rub in a bottom board, vent the mold and roll the drag over. Lay the second half of the pattern over the first making sure that the pegs line up with their mating holes. A good place to gate the mold is at the hub. We used a 1-1/4" round dowel for a sprue. Place a register pin in the bottom of the dowel and drill a matching hole in the center of the hub. This will register the dowel with the hub and help keep it in position as you ram up the cope.

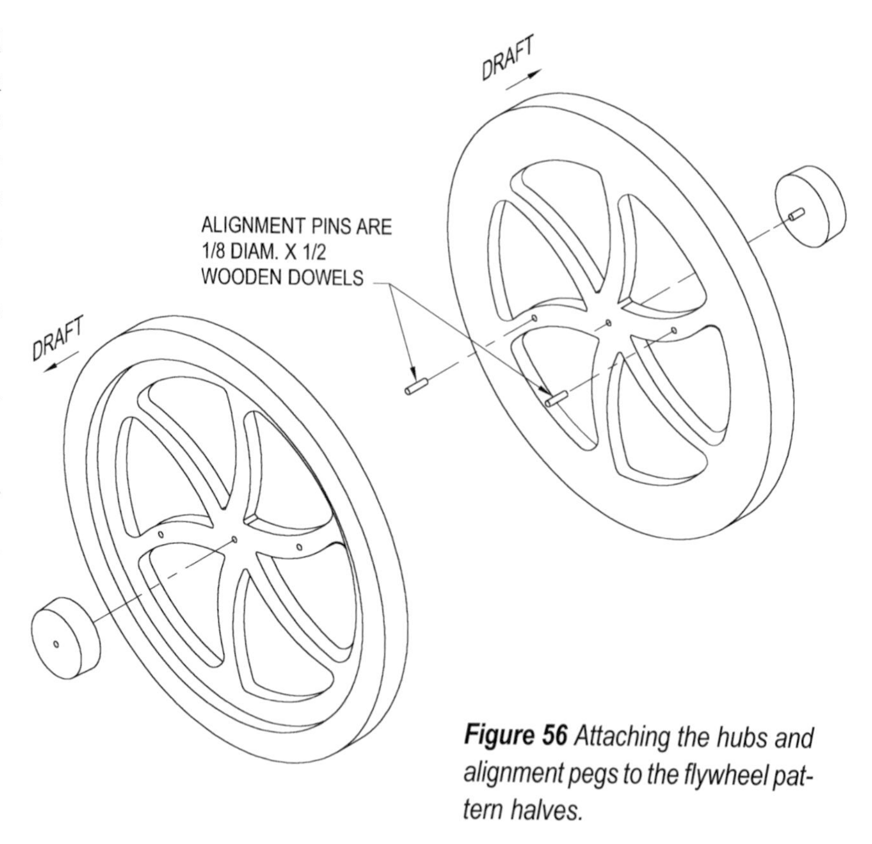

Figure 56 *Attaching the hubs and alignment pegs to the flywheel pattern halves.*

Figure 57 *The flywheel pattern halves.*

Figure 58 *The flywheel pattern assembled.*

Vent the mold, remove the sprue and rub in a bottom board so the cope can be laid down flat to rap and draw the pattern half. Remove both pattern halves from the cope and drag. They should come out easily requiring very little clean up. Reclose the mold and pour. Since this is a large casting, there may be a need to weight the cope to prevent molten metal from escaping at the parting line.

List of materials needed to build the engine . . .

Side panels. Two are required. These can be castings or they can be machined from solid stock. Refer to page 29 for size and pattern detail.

Front & rear panels. One each required. These can be castings or they can be built from solid stock. Refer to page 30-35 for size and pattern detail.

Base. One required. Can be a casting or machined from solid stock. Refer to page 23-25 for size and pattern detail.

Twelve 8-32 x 1/2" hex or socket head machine screws. These are for assembling the main frame.

One 9" flywheel casting. See pages 41-44 for size and pattern detail and 53 & 54 for machining detail.

One tapered hub. Make from a 3" long piece of 1-1/2" diameter c.r.s. round. See figure 69.

Three 10-24 x 3/4" socket head cap screws. For securing the hub insert to the flywheel. See figure 93.

Oscillating arm pivot shaft. Need two each made from 1/2" diameter x 6" long c.r.s round rod. See figure 75.

Two, 1/2" I.D. x 5/8" O.D. x 5/8" long bronze bushings. Crankshaft bushings see figure 81.

Two, 1/4-20 x 1/2" bolts and two 1/4" flat washers. For securing the oscillating arm pivot shafts to the rear panel. See figure 81.

Ignition plate for mounting the points and condenser. Make 1 from a 2-1/2" x 2-1/2" piece of 16 gauge steel. See figure 82.

Two spacers for mounting the ignition plate. Each 1/4" long and made from 1/2" diameter aluminum round rod. See figure 83.

Two, 8-32 x 1/2 machine screws for mounting the ignition plate to the rear panel. See figure 85.

Crank disk. Make one from 1/4" x 3" x 3" c.r.s. flat bar. See figure 86.

Crank shaft. Make one from a piece of 1/2" diameter x 8" long c.r.s. round. See figure 87.

Crank pin. Make one from a piece of 1/2" diameter c.r.s. round. Finished length 1-9/16". See figure 88.

Spacer collar. Make one from a piece of 3/4" diameter x 1/2" long aluminum or steel round. See figure 89.

One 10-24 x 1/2" socket head cap screw with #10 flat washer to secure crank pin to crank disk. See figure 90.

Ignition cam. Make one from a piece of 3/4" diameter c.r.s. round. Finished length 7/8". See figure 92.

One, 1/4-28 x 3/8" set screw for securing the spacer collar. See figure 93.

One, #6-32 x 3/8" set screw for securing the ignition cam. See figure 93.

Water jacket. Make one from a piece of 1" x 2" x 11 gauge wall steel tubing 2-3/8" long. See figure 95.

Material list continued . . .

Cylinder. Make one from a 4-15/16" long piece of 1" black pipe or 1-1/4" diameter cast iron. See figure 94. A good source for cast iron as well as other metal and metal working supplies is, "Nolan Supply", P.O. Box 6289, 111-115 Leo Ave. Syracuse, New York, 13217. Phone 800-736-2204 or see their web page at www.nolansupply.com.

Water jacket base. Make one from a 2" x 3" piece of 10 gauge h.r.s.. See figure 96.

Water jacket cap. Make one from a 1" x 2" piece of 10 gauge h.r.s.. See figure 99.

Spark plug boss. Make one from a piece of 5/8" diameter c.r.s. round. Finished length 5/8". See figure 100.

Exhaust valve boss. Make one from a piece of 1/4" x 3/4" c.r.s. bar stock. Finished length to be 1". See figure 101.

Intake valve boss. Make one from a piece of 1/4" x 3/4" c.r.s. bar stock. Finished length to be 1". See figure 102.

Piston rings. Four are required. Sized to fit a 1-1/8" diameter cylinder. These can be made or you can purchase them through, "Otto Gas Engine Works", 2167 Blue Ball Rd., Elkton, Maryland, 21921-3330. Phone 410-398-7340. See their catalog on the web at, http://www.dol.net/~dave.reed/otto.html

Pistons. Make two from 1-1/4" diameter aluminum round. Finished length of each to be 1-3/8". See figure 128.

Piston rods. Make two from the casting or from solid stock. See figure 130.

Wrist pins. Make two from .3125" c.r.s. round. See figure 139.

Two, 8-32 x 3/8" set screws to secure wrist pins. See figure 139.

Connecting links. Make two from 3/8" x 3/4" aluminum flat bar. Finished length of each link is 2-5/16". See figure 140.

Two, .626" O.D. x 1/2" I.D. x 3/8" bronze bushings. Connecting link bushings. See figure 142.

Link pins. Make 2 from 1/2" diameter c.r.s. round. Final length of each is 29/32". See figure 141.

Oscillating arms. 2 are needed. These can be either castings or they can be constructed and machined from solid stock. See figure 35-46 for pattern and size information and figure 143 for machine specification.

Pivot bushings. 1/2" I.D. x .626" O.D. x 1-1/4" long bronze. Two are needed. See figure 145.

Crank bushings. 1/2" I.D. x .626" O.D. x 1/2" long bronze. Two are needed. See figure 145.

Spacer bushing. Make one from 5/8" diameter aluminum or brass round. Finished length 1/2". See figure 148.

Three 10-24 x 3/8" socket head cap screws each with a #10 flat washer. For securing the linkage. See figure 151.

Material list continued . . .

Piston rod pins. Two are required. Made from 5/16" diameter c.r.s. round. Finished length 3/4". See figure 152.

Two 6-32 x 3/8" set screws. To secure piston rod pin. See figure 152.

Four 8-32 x 3/4" socket head cap screws. For mounting the cylinder to the front panel. See figure 157.

One .030" thick x 2 x 3 paper gasket. Use between cylinder mounting plate and front panel. See figure 157.

Exhaust valve body. Make one from 3/4" x 3/4" x 1" aluminum bar. See figure 158.

Exhaust valve guide. Make one from 1/2" diameter c.r.s. round. Finished length 7/8". See figure 159.

Exhaust valve stem. Make one from 1/8" diameter c.r.s. round. Finished length 1-1/4". See figure 162.

Valve heads. Make two from 7/16" diameter c.r.s. round. Finished length .195". See figure 163.

Valve retainers. Make two from 3/8" diameter c.r.s. round. Finished length 1/8". See figure 170.

Valve springs. Two are needed. Each are cut to a length 4 coils long from a single spring measuring .310" O.D. x .268" I.D. x 1-1/2" long. Wire thickness of the spring is .0210. Such a spring can be purchased at many hardware stores. See figure 169.

Exhaust valve muffler/diverter. Make one from a 3/8" x 3/4" x 1" aluminum block. See figure 171.

Two 6-32 x 5/8" socket head cap screws. For attaching the muffler to the exhaust valve body. See figure 172.

Two 1/16" diameter x 1/4" long retainer pins. Used for securing the valve springs. See figure 172 & 181.

Intake valve body. Make one from a 1/2" x 3/4" x 1" aluminum block. See figure 173.

Intake valve guide. Make one from 1/2" diameter c.r.s. round. Finished length 5/8". See figure 174.

Intake valve cap. Make one from a 1/2" x 3/8" x 1" aluminum block. See figure 177.

Intake valve stem. Make one from 1/8" diameter c.r.s. round. Finished length 1". See figure 178.

.030" thick paper gasket material. Size 1/2" x 1". Used between the intake valve cap and the intake body. See figure 181.

Two 6-32 x 1/4" socket head cap screws. For securing the intake valve cap to the intake valve body. See figure 181.

Two 6-32 x 7/8" socket head cap screws. For securing the exhaust valve assembly to the cylinder. See figure 182.

Two 6-32 x 5/8" socket head cap screws. For securing the intake valve assembly to the cylinder. See figure 182.

Carburetor main body. Make one from 1/2" diameter aluminum or brass round. Finished length 1-1/8". See figure 183.

Material list continued . . .

Air intake orifice. Made one from 1/2" diameter aluminum or brass round. Finished length 3/16". See figure 184.

Jet. Make one from 3/16" brass hex rod. Finished length 1/2". See figure 186.

One 6-32 x 3/4" brass screw. For needle valve. See figure 187.

One 3/8" diameter x 3/32" long brass knurled nut. For needle valve. See figure 187.

One #5 sharp sewing needle. See figure 187.

One spring from a retractable ball point ink pen. See figure 188.

One set, ignition points and condenser from a late 70's ford V8. See figure 189.

Three 6-32 x 1/2" machine screws. To secure points & condenser. See figure 189.

One spark plug. Either NGK #CM6 or a Champion #Y82. See figure 190.

Gas tank. Make one from a sheet of .025 thick copper, brass or sheet metal measuring 2-1/4" x 5-1/2". See figure 191

Gas tank end caps. Make two. Each is made from a 1-3/4" diameter piece of .025 thick copper, brass or sheet metal. See figures 194-197

Gas tank outlet fitting. Make one from a piece of 3/8" diameter brass round. Finished length 1/2". See figure 199.

Gas tank inlet fitting. Make one from a piece of 1/2" diameter brass round. Finished length 3/8". See figure 200.

Gas cap. Make one from a piece of 5/8" diameter aluminum or brass round. Finished length 1/2". See figure 202.

Fuel tank adaptor. Make one from a piece of 3/8" diameter brass round. Finished length 1/2". See figure 203.

Fuel pickup tube. Make one from a piece of 1/8" O.D. brass tubing 1-3/8" long. See figure 203.

Gas line hose adapter. Make one from a piece of 1/4" brass hex. Finished length 1/2". See figure 204.

One .1565" diameter ball bearing. Fuel check ball. See figure 205.

Approximately 10" of 1/8" I.D. vinyl hose. Fuel line. See figure 206.

Tie bar. Make 1 from a length of 1/2" x 3/4" aluminum flat bar. Length to be determined during construction. See figure 207.

1 ignition coil. Can be either a lawnmower coil or an automotive coil. See figure 208.

1 battery. Depending on the ignition coil used, the battery can be either 6 or 12 volts. See figure 208.

Building the Atkinson Differential engine . . .

It is likely that the oscillating arms and other moving parts will appear to be the greatest challenge in this engine project. In fact the only truly demanding part of the job is to assemble the five simple castings that comprise the main frame and the water reservoir. But though demanding accuracy, the work is not difficult when taken one step at a time. The objective is to ensure that the assembly is watertight and that the boring operations are aligned front-to-back, parallel to each other and exactly perpendicular to the vertical axis of the main frame. To achieve this, the mating surfaces must be 90 degrees at the corners and nicely flat. The best idea appears to be to assemble the main frame as carefully as possible and then disassemble it for the front to back boring operations if your equipment is not large enough.

Prepare the end panels . . .

The only required work is to bring them to size, making the edges flat and perpendicular to the outside surface and making the vertical edges perpendicular to the bottom edges. The top edge needs no work and the outside surface will look best as cast.

Figure 59 *Square the side panels.*

Prepare the front & back panels . . .

Just file away evidence of the gates and chase the smooth surface with coarse emery and hammer to make it look as cast. Lightly file the inside surfaces to knock off any "flash" and "nubbins". Then mill the bottom edge flat and perpendicular to the inside surface and ensure that the bottom edge is perpendicular to the vertical edges. The bottom edge of both panels can be milled at the same time. Find and use the center mark on one of the bosses to determine how much can be milled off the bottom edge. See figure 60 for setup and the drawing in figure 62 for the required height from the bottom edge of the front panel to center of bosses.

Figure 60 *Position the front & rear panels back to back. Align the outside edges as close as possible. Then clamp them as a pair to the mill table and mill the bottom edge of both panels at the same time. When finished, the bottom edge of each panel should be flat*

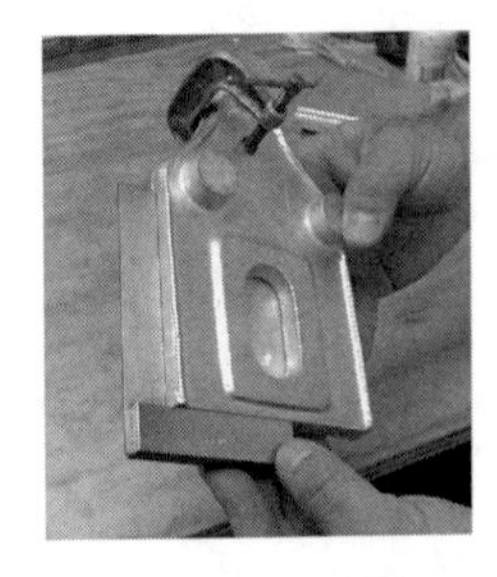

Figure 61 & 61A *Checking the front and back panels with a square to ensure bottom edge is flat and perpendicular to the inside surface and perpendicular to the vertical edges.*

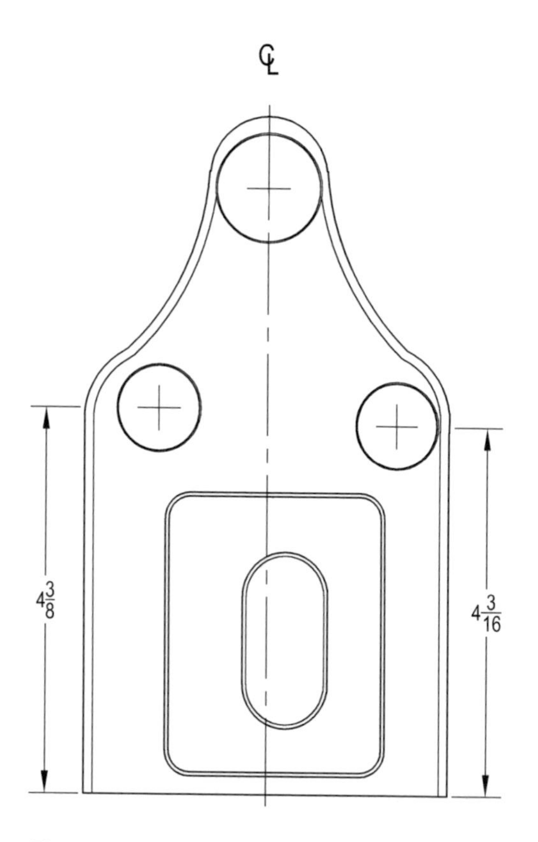

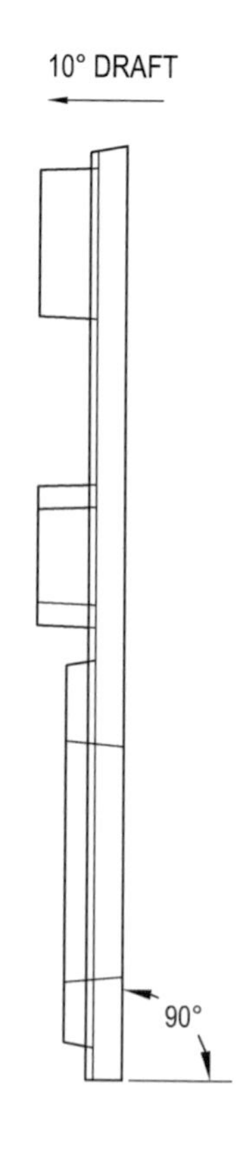

Figure 62 After milling the bottom edge of the panels the height should measure from the bottom edge of the panel to the boss centers as shown in the drawing.

Other areas of the base that may require attention are those areas where the main frame will rest and only then if the surface is extremely rough. RTV or other sealing material will take care of most minor irregularities. File the gate area flush with the adjacent area. Then lay coarse emery cloth over any smooth spots left by the file and tap with a hammer to give the surface that "as cast" look. If you are using solid stock you can mill the flange around the edge or not as you choose. And you can mill the cavity on the underside, mill a channel or simply counter bore for the bolt heads.

Prepare the base . . .

If your are using the prescribed casting, the only machining that may be needed on the base is to mill or file the bottom of the legs so the base will rest evenly on a flat surface.

Figure 63 The base

Figure 64 Test alignment. When the parts fit nicely together you can begin to drill and tap for assembly.

Assemble the main frame . . .

There will be twelve pairs of holes to drill and tap for the initial assembly. 8-32 machine screws are adequate. While slotted or Phillips head screws will work, those that are visible should be hex or socket head for the most authentic appearance. If you do not find a ready source for the hex head machine screws you can make them on the lathe using steel or brass hex stock and threading them with a tail stock die holder. It's worth the effort. The front frame panel shown in figure 65 is layed out carefully, center punched and drilled 11/64 in four places. Its base flange is drilled #29 and tapped 8-32 on center and 1/8" from the front edge.

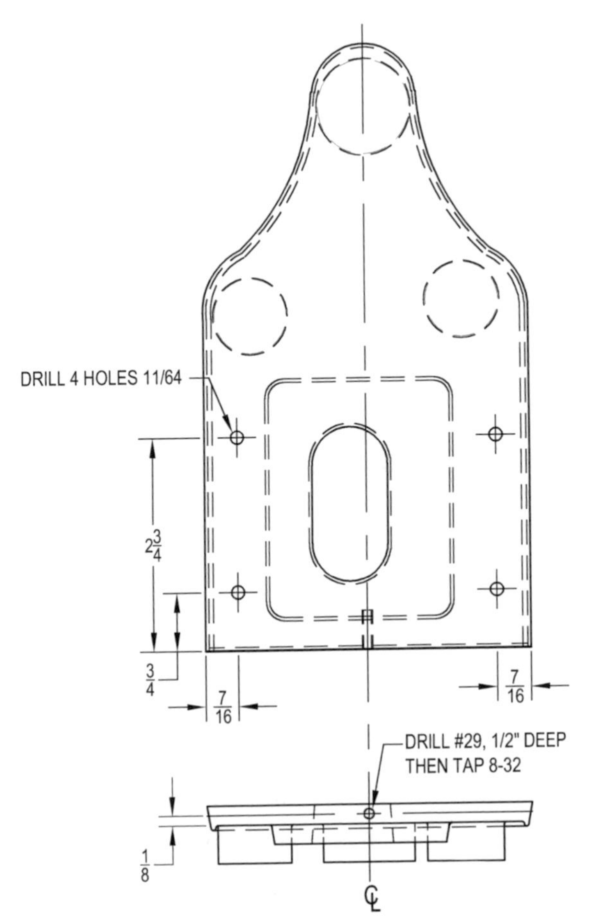

Figure 65 *Front panel. Drill the four, 11/64" holes then drill and tap the bottom edge on center for 8-32 thread.*

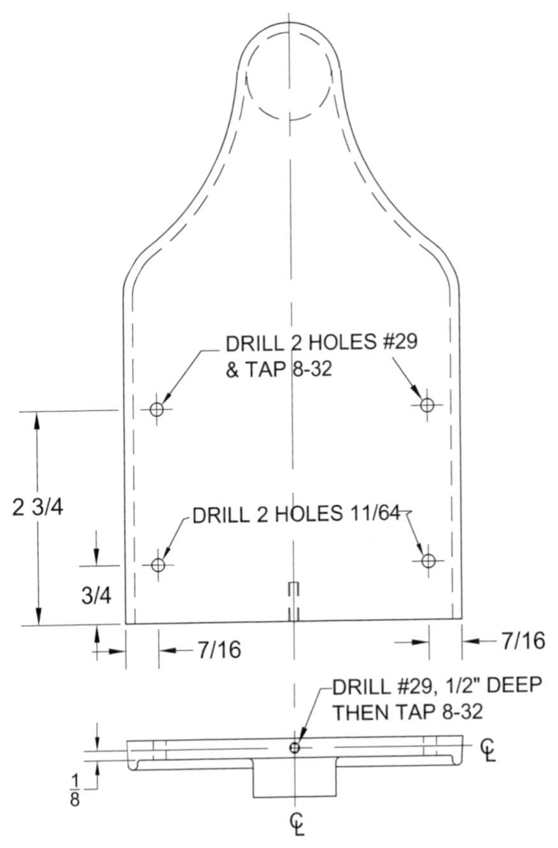

Figure 66 *Rear panel. Drill two, 11/64" holes and drill and tap the two #8-32 holes. Then drill and tap the bottom edge on center for 8-32 thread.*

The rear panel shown in figure 66 is layed out exactly like the front panel, but two of the four holes are drilled #29 and tapped 8-32 as shown. These will be used to hold front and back panels together for boring. Later they will be enlarged to 11/64 for final assembly.

The side panels shown in figure 67 are drilled #29 and tapped 8-32 in five places to correspond with the 11/64" holes in the front, back and base panels. Be thoughtful and careful as you layout, drill and tap. It's a good idea to use the drill press to start the tap by hand to assure alignment and avoid breaking taps. When these parts are drilled and tapped assemble them with 8-32 x 1/2" screws and check for squareness and hole alignment. 11/64" holes are large enough for clearance if the parts are true and accurately layed out. But they can be enlarged if necessary. Leave the screws loose until all are well started. If alignment problems are encountered it will be better to enlarge all 11/64" holes to 3/16" rather than trying to elongate one hole. And you may have to go to 13/64" or larger if you have not yet perfected your layout skills.

The fundamental rule is: *"Layout from a common reference base, center punch carefully, use a sharp drill in the drill press and start the tap straight."* The most common cause of tap breaking is starting them crooked.

The base panel shown in figure 68 should require no machining unless it does not lay true flat. Four 13/64" holes are drilled in the base for mounting the engine. To locate the position the main frame on the base, carefully layout horizontal and vertical center lines on the raised pad of the base. Then center the assembled main frame on the base and scribe a line along its outer edges to mark its location. Layout the 4 hole centers in the base 1/8" from the the scribed lines to correspond with the 8-32 holes in the bottom of the main frame. Center punch and drill 13/64" in four places. Also, locate, drill and tap a 1/4-20 hole in each foot pad as shown in the drawing.

THE INSIDE DASHED LINE REPRESENTS THE SCRIBED LOCATION OF THE ASSEMBLED MAIN FRAME. SEE TEXT.

DRILL 4 HOLES 13/64

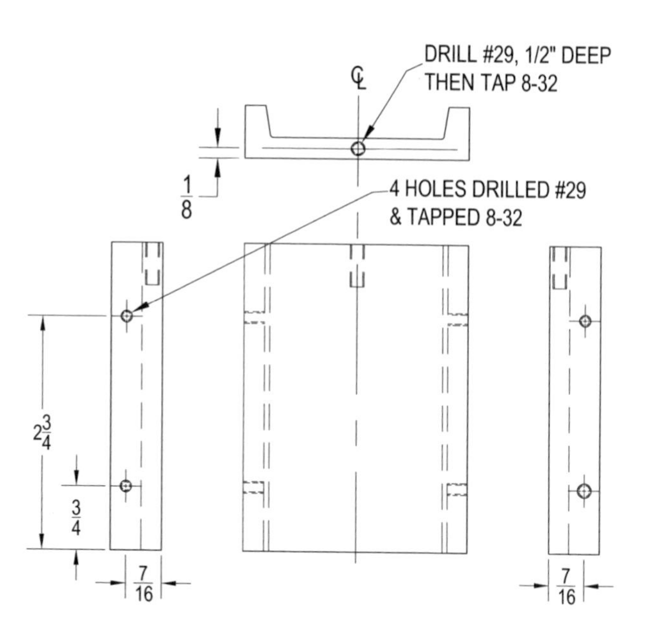

Figure 67 *Side panels. Prepare two as shown.*

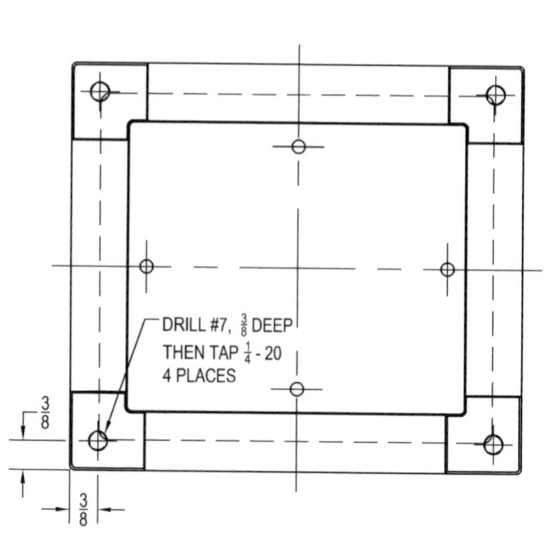

Figure 68 *Locating the main frame mounting holes in the base panel.*

Now you can screw the base to the main frame with four, 8-32 x 1/2" screws. At this stage we are concerned only with a well aligned assembly. Enlarged holes will be covered by screw heads, or even flat washers if need be. On final assembly, we'll use polyester body putty or RTV sealer on the joints to make them water tight. A coarse belt on the belt sander will do an effective job on the bottom edges of the frame and the top surface of the base.

Prepare the flywheel . . .

Our decision when we built our engine was to have the back curve of the spokes face the direction of rotation. I have seen it both ways and I am not sure if it really matters or not. Anyway, that's the basis we used in determining the front and back of the flywheel. The sides are labeled in figure 70.

The front side of the hub has a tapered bore. A split tapered hub insert fits into the tapered bore. Then as the tapered hub insert is tightened and pulled into the hub of the flywheel it clamps down on the crankshaft securing the flywheel to the shaft. This method of holding the flywheel on the shaft works very well and is certainly much better than set screws which always seem to work loose. Another nice feature is the ease in which the flywheel can be removed. It's simply a matter of removing the bolts and screwing one into each of the threaded holes in the insert. As the bolts are tightened they push against the flywheel hub and force the insert out loosening the flywheel.

To make the tapered hub insert shown in figure 69, chuck a 3" length of 1-1/2" diameter c.r.s. round in a 3-jaw chuck. Face off, center drill, to near size and ream .501" 1-1/4" deep. Turn the 7/8" shoulder 7/8" back. Adjust the compound rest and turn the 7 degree taper. Part off the insert 1-1/8" long. Layout and drill the three 3/16" mounting holes 9/16" from center and 120 degrees apart. One way to divide the circle in three equal parts is to wrap a piece of masking tape around the outside diameter. Trim the tape to fit exactly. Remove the tape and divide it into 3 equal sections marking each section. Rewrap the tape around the outside diameter. Use a tri-square and the marks on the tape as a reference and mark the location of each of the three holes.

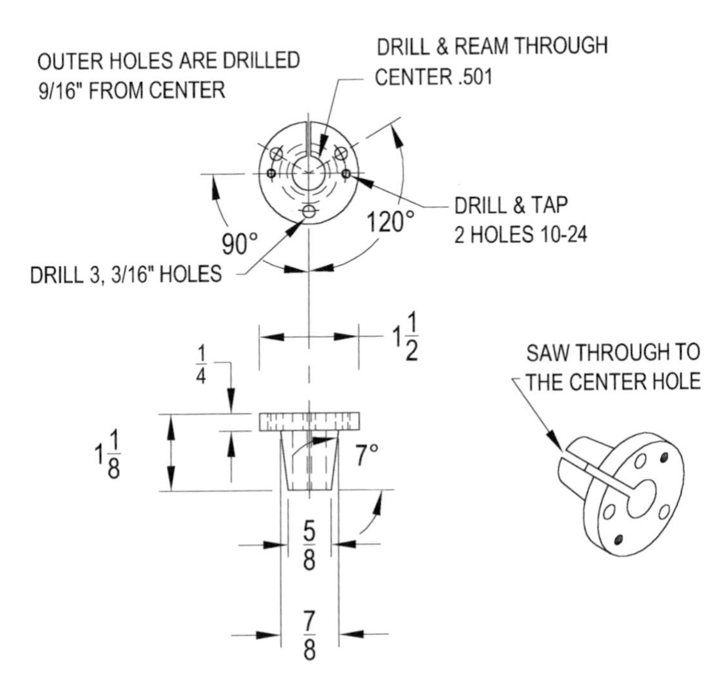

Figure 69 *The flywheel hub insert.. Make 1 from 1-1/2" diameter c.r.s. round.*

Saw through one side of the hub insert where shown in the figure. Locate, drill and tap the two threaded holes in the hub insert 10-24. They are located on the center line and perpendicular to the saw slit.

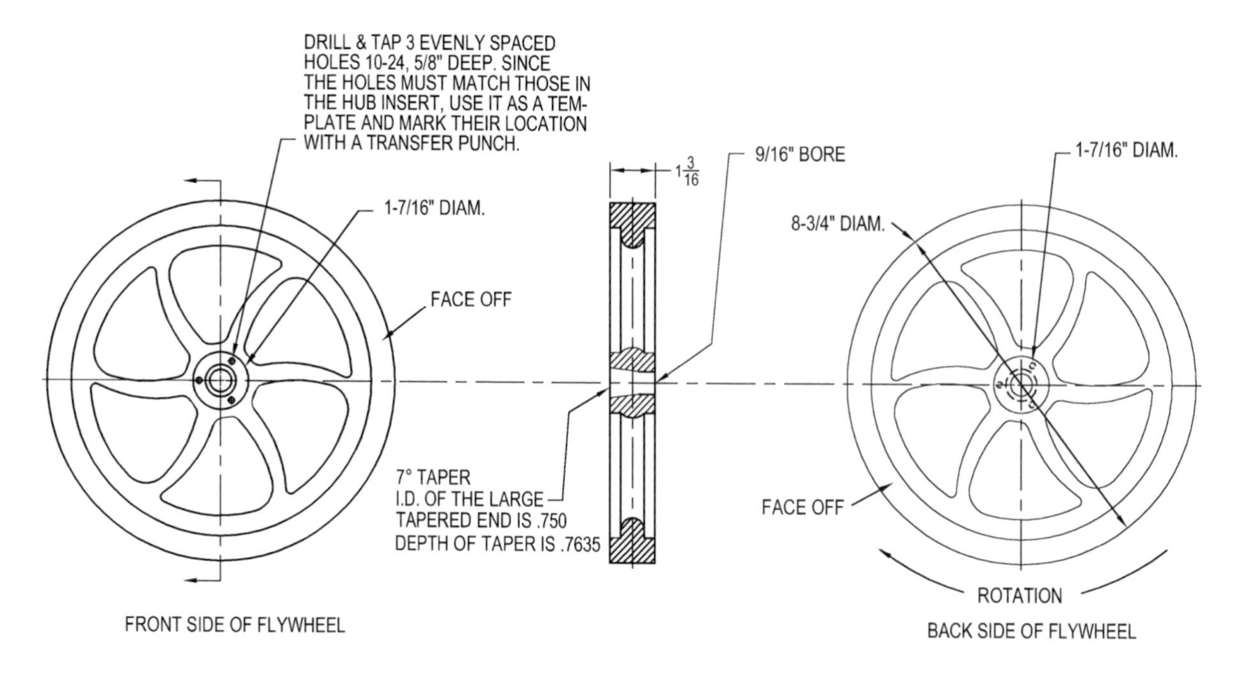

Figure 70 *Machining detail for the flywheel.*

The finished dimensions for the flywheel are shown in figure 70. Begin the machine work by chucking the front side of the flywheel in a 4-jaw chuck. Center it as close as possible. Face off and turn the outside diameter of the hub.

Remove the flywheel and rechuck it from the rear centering it in the chuck. Then face off and turn the outside diameter of the hub. Bore through the center of the hub to 9/16" diameter then bore the 7 degree taper.

To locate the bolt pattern for the hub insert on the face of the hub on the flywheel, use the hub insert as a template. Mark the hole locations with a transfer punch. Drill and tap them 10-24, 5/8" deep.

To machine the outside diameter of the flywheel it is best to mount the wheel on an arbor for between the centers work. To prepare the arbor, chuck a 1/2" shaft in the lathe with about 5" protruding. Center drill the end to support the shaft with a tail center. Then mount the flywheel on the arbor using the tapered hub insert. Continue to support the arbor with the tail center while machining the O.D. and sides of the flywheel. To face off the outer rim of the back side of the flywheel you will have to remount the flywheel on the arbor.

Figure 71 *Finished flywheel & hub insert.*

Boring for crankshaft bearings and pivot shafts . . .

Our primary concern is to have the crankshaft and oscillating arm pivots perpendicular to the inside surface of the front frame and parallel to each other. Unless you have a very large mill or lathe, this work will require some scheming. If you have a large mill or lathe with required tooling no discussion is required. But if you are working with only a drill press or a small mill/drill and a small lathe there are some tricks of the trade that will aid you.

By assembling the panels back to back and with careful layout, center punching and step drilling, you can bring the bores to near size on a good drill press. Then we used modified chucking reamers to finish them to size so they will be aligned front to back and parallel.

You will need 3 reamers to finish the engine but only two will be modified, the wrist pin and link pin are 5/16" and for that you need a reamer .001" oversize or .3135. The crank bearings and oscillating arm bearings are 1/2" and for that you will need a reamer .001" over size or .501". The crank bushings and oscillating arm bushings are 5/8" OD and for that you need a standard 5/8" reamer. The modification to the 1/2" and 5/8" reamers include squaring the shank end and preparing guide bushings for the shanks.

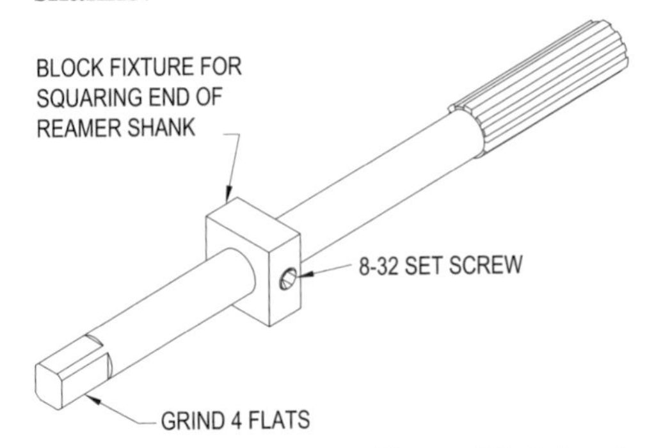

Figure 72 Setup for grinding flats on the end of the reamer.

Figure 73 Grinding flats on the end of the reamer shaft. Notice, block fixture resting on grinder table.

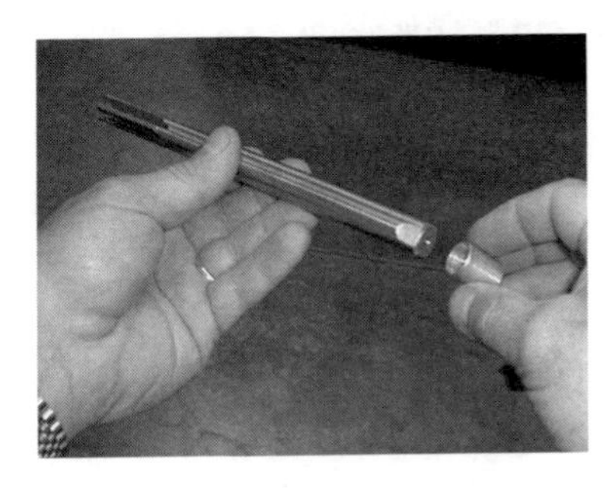

Figure 74 Alignment bushing and reamer.

It's presumed that you will use chucking reamers because hand reamers are not easily available and are quite expensive if found. It is an easy job to fit a square on the end for the tap wrench. The shanks of chucking reamers are typically 1/16" smaller than the nominal size so the 1/2" reamer requires a block drilled 7/16" and the 5/8" reamer requires a block drilled 9/16".

The 1/2" and 5/8" reamers will each require guide bushings that fit the shank closely and slip into the nominal size bore.

With a short squared shank and a guide bushing you have converted your cheap chucking reamers into the practical equivalent of expensive self aligning hand reamers.

You will need two oscillating arm pivot shafts in the sequence so make them now from 1/2" cold rolled or drill rod. See

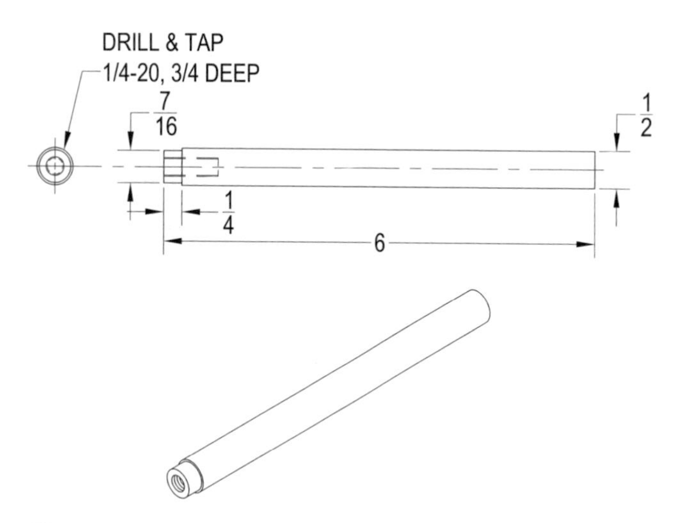

Figure 75 *Oscillating arm pivot shaft. Make two from 1/2" diameter c.r.s.*

figure 75. Simply drill and tap one end of each for 1/4"-20 threads, 3/4" deep and reduce the end to 7/16" diameter 1/4" back.

Disassemble the frame assembly and carefully mark and punch the bore locations in the three bosses in the front panel where shown in the figure 76. The bore locations need to be located as close as possible to the dimension shown in the drawing. In some cases the holes through the bosses may not be exactly in the center of the boss. That's not important, but the dimensional location is.

Assemble the panels back to back with two bars of 1/4" or heavier cold

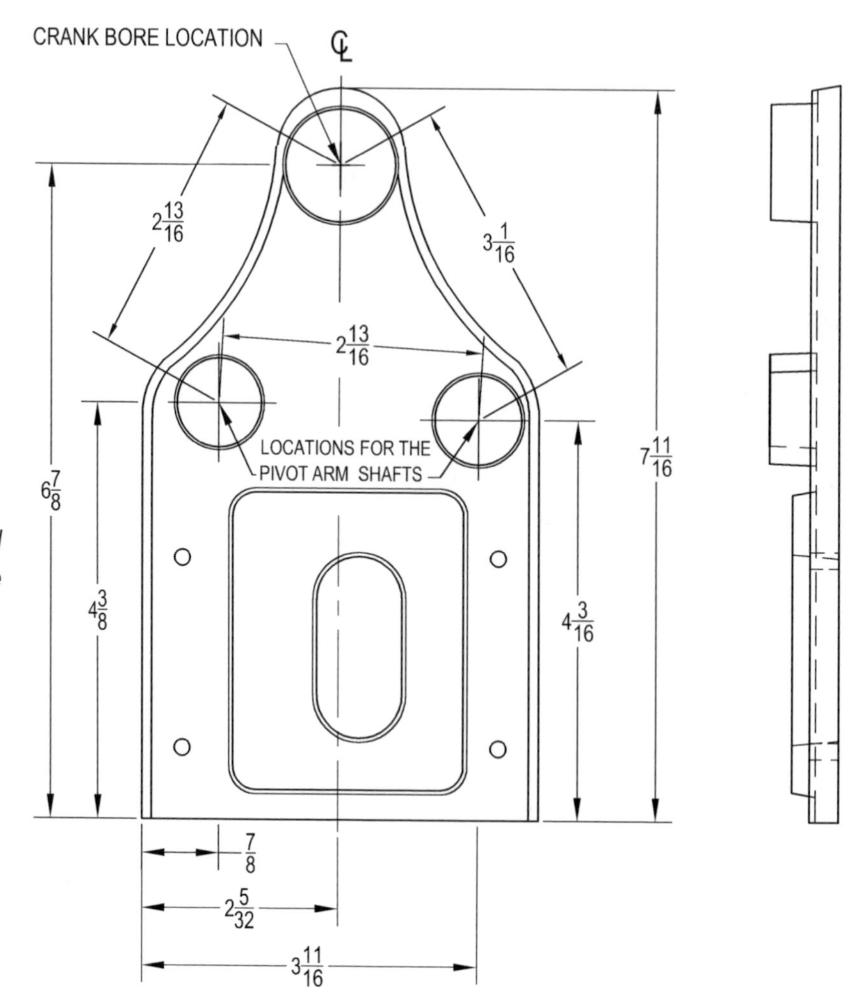

Figure 76 *Drawing of front panel showing the locations for the crankshaft and pivot arm bores.*

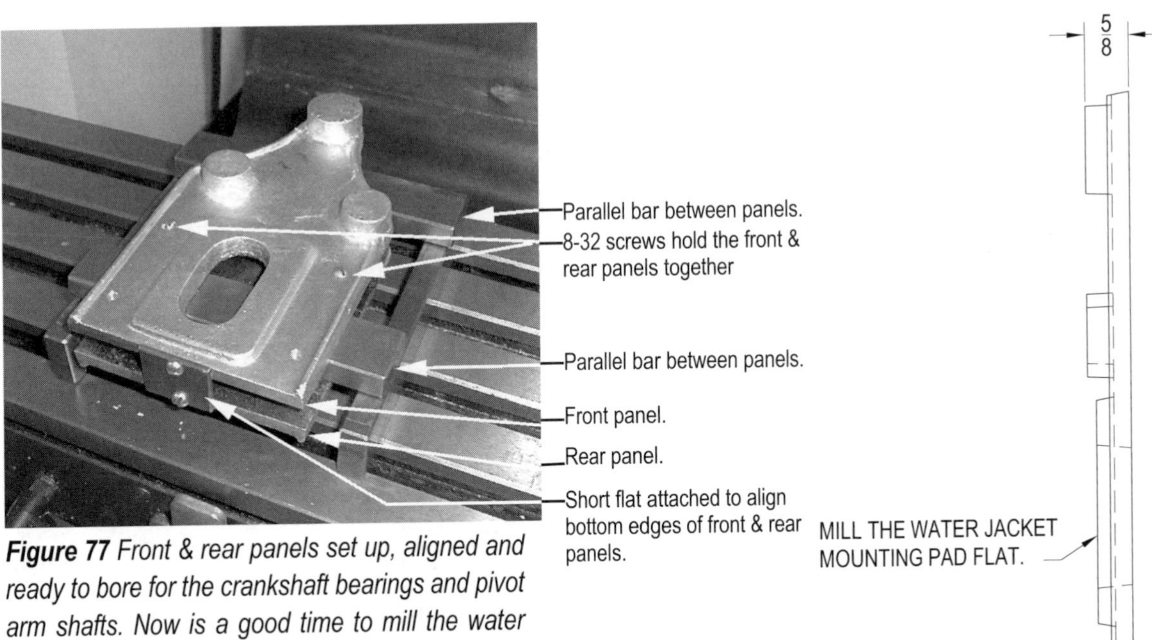

Figure 77 *Front & rear panels set up, aligned and ready to bore for the crankshaft bearings and pivot arm shafts. Now is a good time to mill the water jacket mounting pad flat and also to mill the bosses. Dimension of bosses from back of panel to front of boss to be 5/8". See side view drawing located at the right of the photo above.*

Figure 78

rolled flats 6" long sandwiched between them. Use a short flat drilled to align the bottom edge. See figure 77.

The flats will hold the members aligned and parallel and a pair of 3/4" parallels will support the assembly on the drill table. Now you can begin drilling with an 1/8" drill and increase in practical increments until all three bores are 7/16". Then increase in increments until the crank bore is 9/16". At this stage the shanks of the reamers should slip nicely through the bores. But 1/16" is too big a bite for accurate work with a hand reamer. So you leave the back panel at 1/16" undersize to support the reamer shank and enlarge the 2 pivot bores in the front panel to 31/64" and the crank bore to 39/64". But we'll assemble the main frame with its side panels and base before we drive the reamers through.

Figure 79 *In this photo, one pivot shaft has been installed and we're getting ready to ream for other.*

The 8-32 tapped holes in the back panel can now be enlarged to 11/ 64". Assemble the back and side panels to each other and the base and slip the shank of the .501" reamer into either of the 7/16" holes in the back panel. Assemble the front panel to the side panels and the base and check for squareness before driving the reamer through the front panel. Withdraw the reamer and remove the front panel to move the reamer to the second bore in the back panel. Reassemble the front

Figure 80 *Reaming for the crankshaft.*

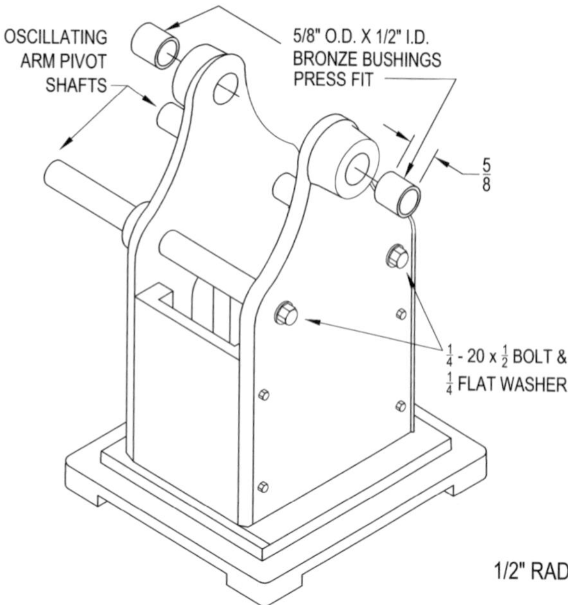

Figure 81 *Location of the crank bushings in front & rear panel.*

panel and install one oscillating arm shaft in the reamed bore before reaming the second bore. Draw the shaft up snug with a 1/4"-20 x 1/2" capscrew with a flat washer. Drive the reamer through the second bore and remove the front panel to slip the 5/8" reamer into the 9/16" bore in the back panel. Reinstall the front panel and install both oscillating arm shafts before you drive the 5/8" reamer through the front panel. When the front panel is reamed 5/8", remove the back panel and enlarge the bore to 39/64" on the drill press. Reinstall the back panel with both oscillating arm shafts snugged up and ream the back bore to 5/8" using the shank bushing to align the reamer. If the process is done carefully a 1/2" crank shaft will fit nicely through the bushings in both panels and the crank and arm shafts will be parallel.

Press a 5/8" O.D. x 1/2" I.D. bronze bushing into the crankshaft boss of the front and then the rear panel.

You might want to take time to paint the base and main frame panels now. The paint can be drying while you make the next few parts.

The ignition plate . . .

Now will be the best time to make and install the ignition plate. The points and condenser can be mounted on the plate now or later as you choose. Look ahead to figure 197 for information on mounting the

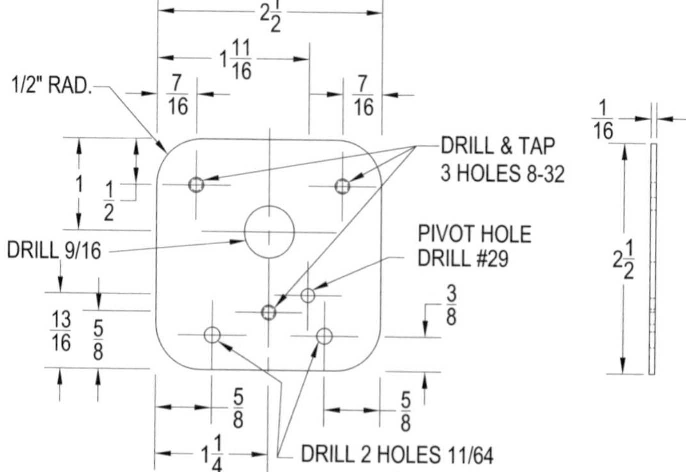

Figure 82 *Ignition plate. Make 1 from 16 gauge steel.*

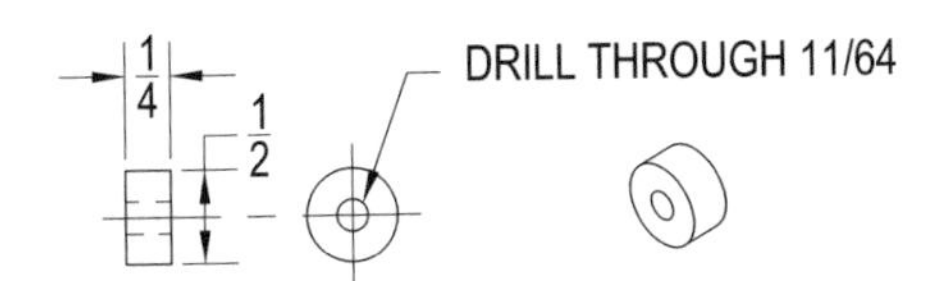

Figure 83 *Spacer for ignition plate. Make two from 1/2" diameter aluminum or steel round rod.*

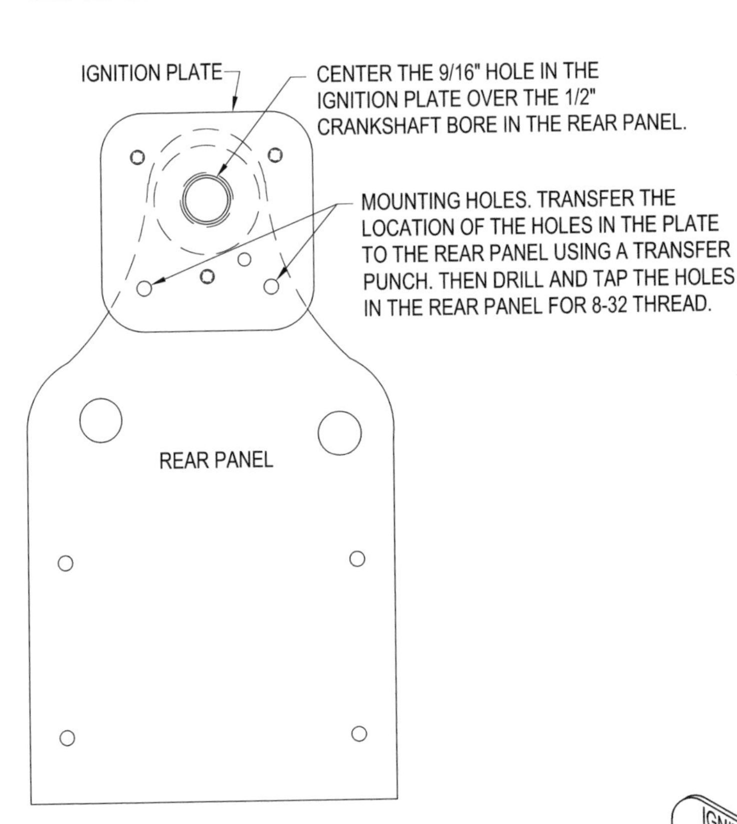

Figure 84 *locating the ignition plate mounting holes in the rear panel.*

ignition points. The plate is made from a piece of 2-1/2" x 2-1/2" x 16 gauge steel. For reference, 16 gauge steel measures about 1/16" thick. The layout for the ignition plate is shown in figure 82.

Make the two spacer pads shown in figure 83. They are used between the ignition plate and rear panel as shown in figure 85.

The ignition plate will mount to the back side of the rear panel. The best way to locate the 8-32 threaded holes in the rear panel for mounting the ignition plate is to use the mounting holes that exist in the plate as a template. Position the plate on the backside of the rear panel as shown in figure 84. You want the 9/16" hole in the ignition plate centered over the 1/2" crankshaft hole in the rear panel. When the plate is in position, secure it to the rear panel with a "C" clamp. Ensure that the sides of the ignition plate are parallel with the sides of the rear panel before tightening the clamp. Use a transfer punch to mark the location of the mounting holes in the rear panel. Remove the plate, drill each hole #29, then tap 8-32. Give the ignition plate a coat of paint, then mount it to the rear panel using the spacers from figure 83 positioned between the panel and the plate. Secure with 8-32 x 1/2" machine screws as shown in figure 85.

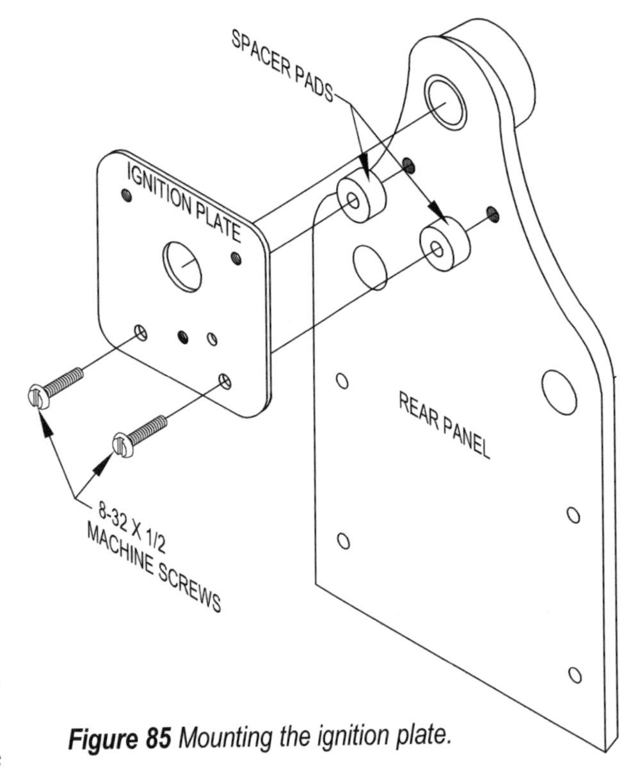

Figure 85 *Mounting the ignition plate.*

Crank shaft assembly . . .

The crank shaft assembly consists of the crankshaft, crank disk, crank pin and flywheel.

Make the crank disk shown in figure 86 from a piece of 3" x 3" x 1/4" c.r.s. flat bar. Locate drill and ream the .501" hole in the center of disk blank. Then locate, drill and ream the .375" hole in the blank.

Cut an 8" length of 1/2" diameter c.r.s. round rod for the crankshaft. Insert one end of the crankshaft into the .501" hole in the center of the crank disk blank as shown in in figure 87. The end of the crank shaft should be flush with the front surface of the crank disk. When positioned, braze the crank shaft to the crank disk.

Chuck the end of the crankshaft in the lathe to face off then turn the outside diameter of the crank disk blank to 3".

Make the crank pin from a 1-9/16" long piece of 1/2" c.r.s. round rod as shown in figure 88. Turn the .3760" end for a press fit into the crank disk. Drill and tap both ends for 10-24 thread.

Next make the spacer collar as shown in figure 89.

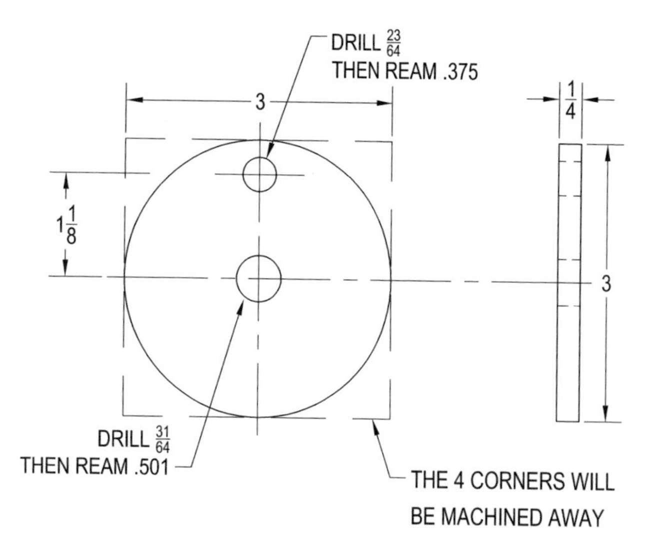

Figure 86 *Crank disk. Make 1 from a piece of 1/4" x 3" x 3" c.r.s. flat bar.*

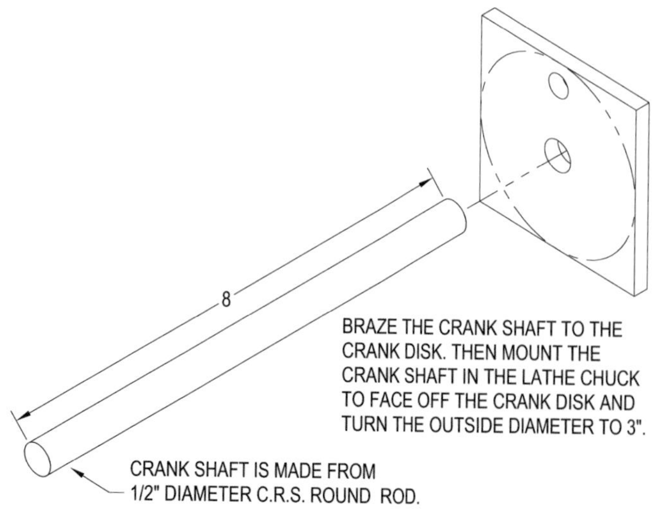

Figure 87 *Brazing the crank shaft to the crank disk. The crank shaft is made from a 8" length of 1/2" diameter c.r.s. round rod.*

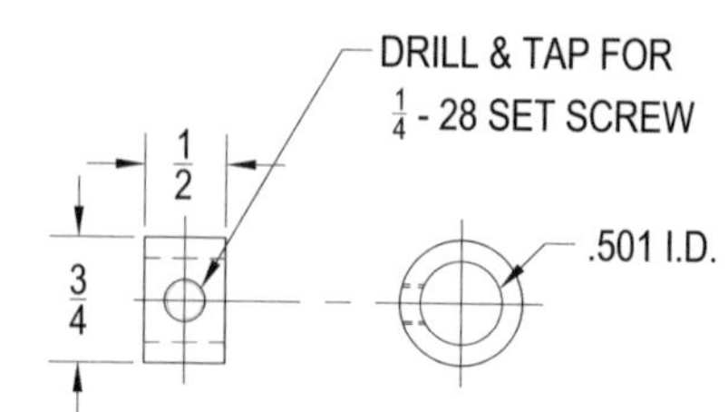

Figure 89 *Spacer collar. Make 1 from 3/4" diam. aluminum or steel round.*

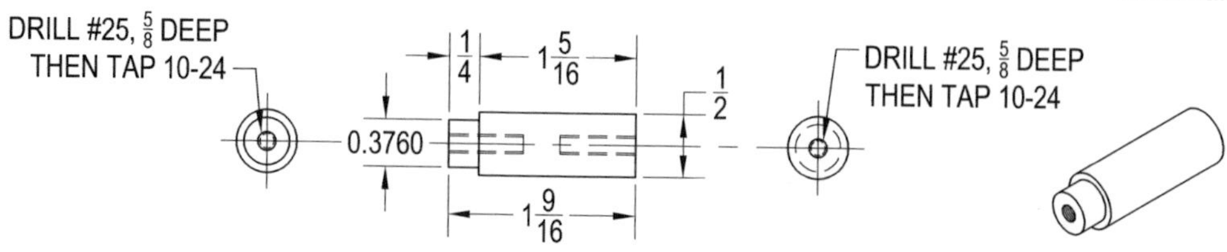

Figure 88 *Crank pin. Make 1 from a piece of 1/2" diameter c.r.s round.*

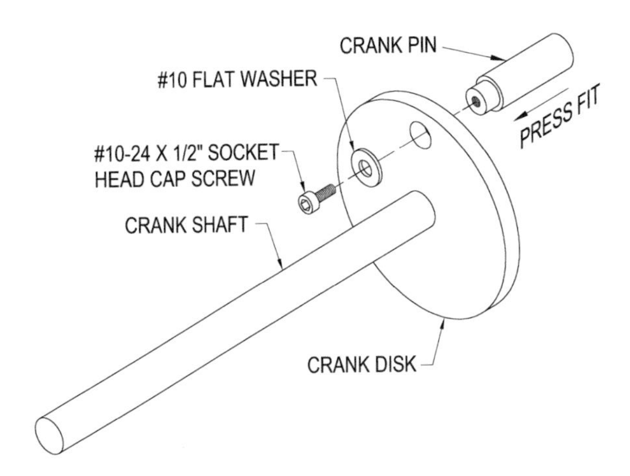

Figure 90 *Installing the crank pin.*

Press the .3760" end of the crank pin into the .375" hole in the crank disk. Further secure the crank pin to the disk with a 10-24 x 1/2" cap screw and #10 flat washer. The cap screw will help ensure the crank pin does not work loose. Figure 91 shows a side view of the completed crank shaft assembly.

Make the ignition cam as shown in figure 92 from a piece of 3/4" diameter c.r.s. round.

Before final assembly, drill 1/16" oil holes through to the crank bore on both the front and rear panels. Countersink each hole to provide a cup for the oil. See figure 93.

Assemble the crank shaft, ignition cam and flywheel in the main frame of the engine as shown in figure 93.

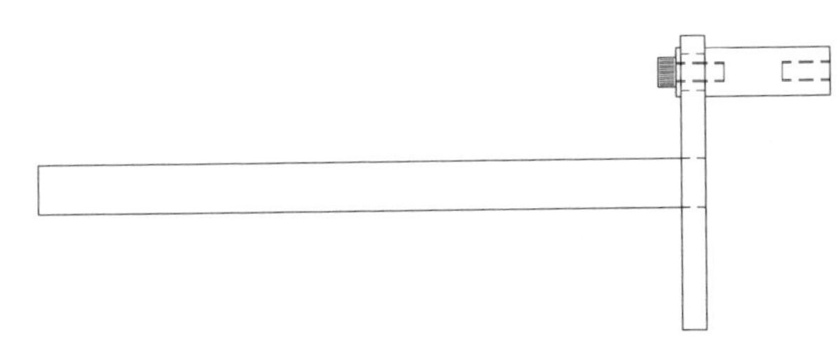

Figure 91 *Side view of crank shaft assembly.*

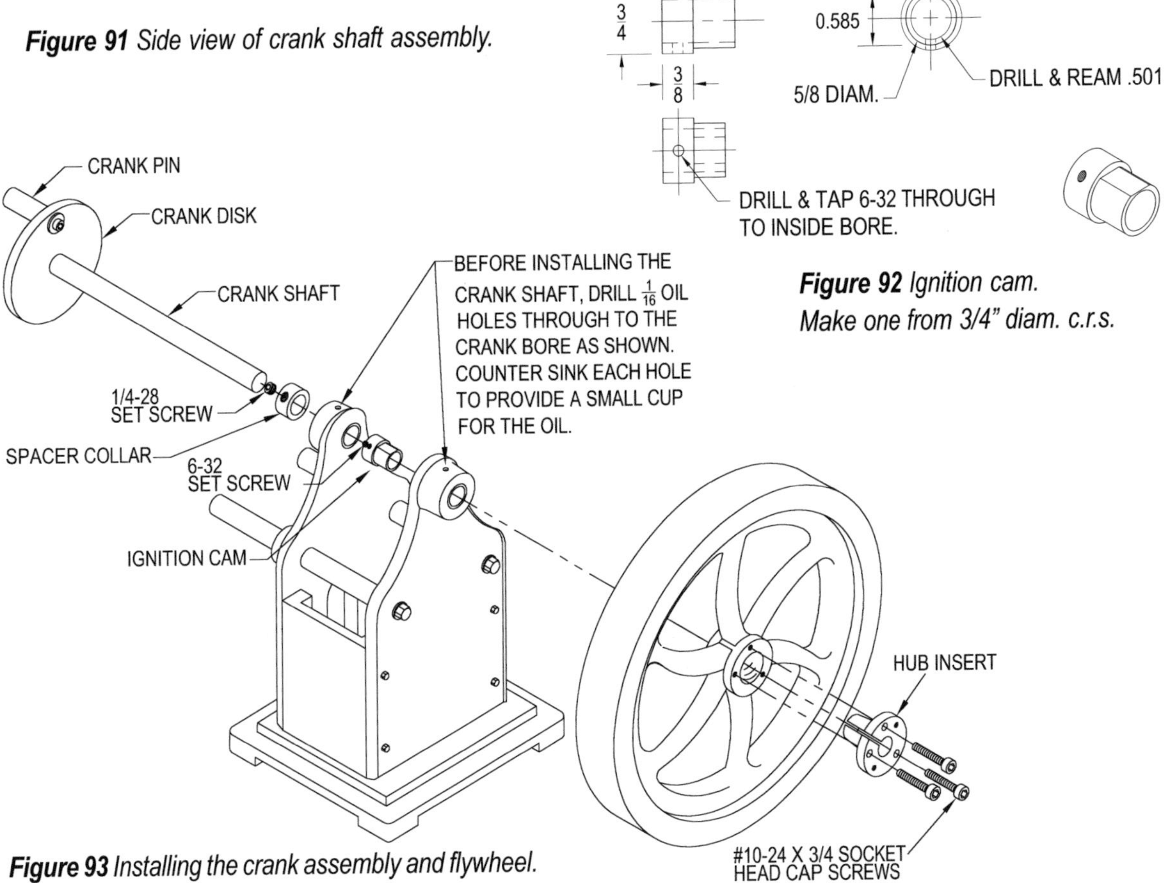

Figure 92 *Ignition cam.*
Make one from 3/4" diam. c.r.s.

Figure 93 *Installing the crank assembly and flywheel.*

The cylinder water jacket assembly . . .

It was preferred to use a casting for the water jacket but that approach proved too demanding so we fell back on the easier method of welding and brazing the assembly instead.

As for the cylinder, cast iron is generally preferred for engine cylinders, but common 1" black iron pipe will also work. And if you use pipe, you will find it is easily worked into a 1-1/8" cylinder and readily available almost everywhere.

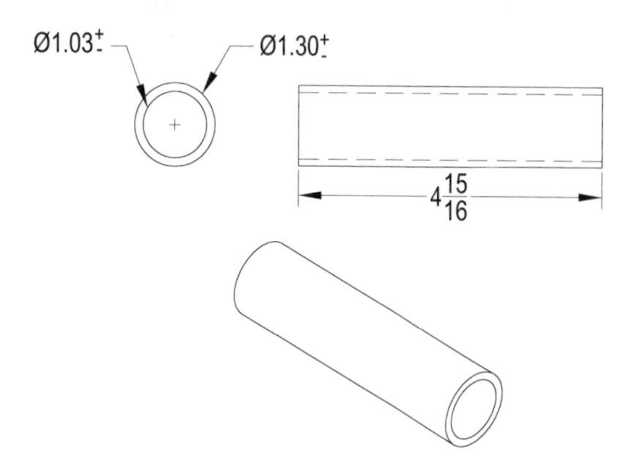

Figure 94 Cylinder. Cut one to length from 1" black iron pipe or 1-1/4" diameter cast iron.

The weight of the material used for the water jacket shown in figure 95 is not critical. But since the cylinder wall thickness is near 1/8" it will be best to use stock near that size. The jacket sleeve is the equivalent of 1" x 2" rectangular steel tubing. 11 gauge wall steel tubing is .120" thick and ideal. But if rectangular tubing is not readily available you can weld up a tube from 1/8" flat stock.

The water jacket can be bored on the lathe or on the mill/drill depending upon available equipment. Options may include using the 4-jaw chuck, an angle plate or a jig. The objective is to bore on the center line 1-5/16" from the mounting surface.

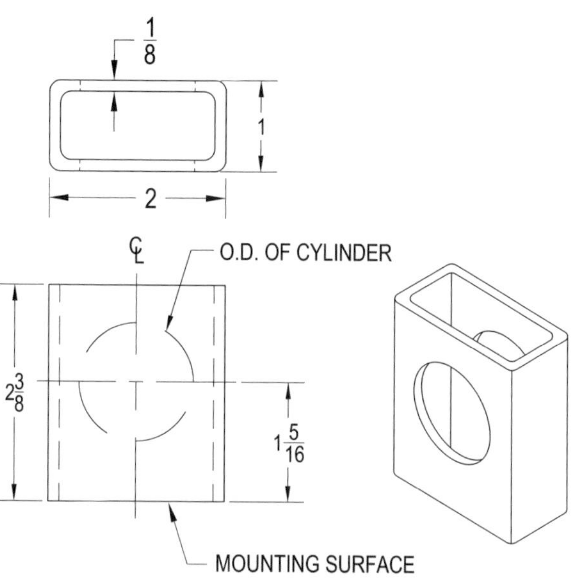

Figure 95 Water jacket. Cut to length from 1" x 2"" x 2-3/8", 11 gauge wall rectangular steel tubing. Bore through to match the outside diameter of the cylinder.

The base for the water jacket shown in figure 96 is also made of 1/8" flat steel prepared as shown. 10 gauge h.r.s. is near 1/8" thick and works fine. The resulting rectangular is welded or brazed to the water jacket sleeve and a 10 gauge cap is welded or brazed to the top to finish it up as shown in figures 97 and 99.

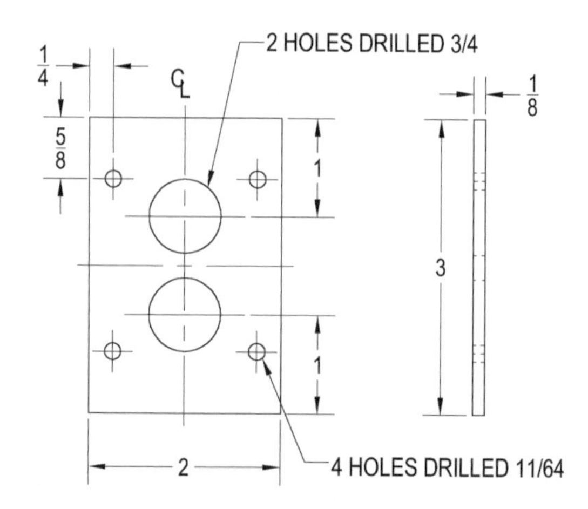

Figure 96 Water jacket base plate. Make 1 from 10 gauge h.r.s.

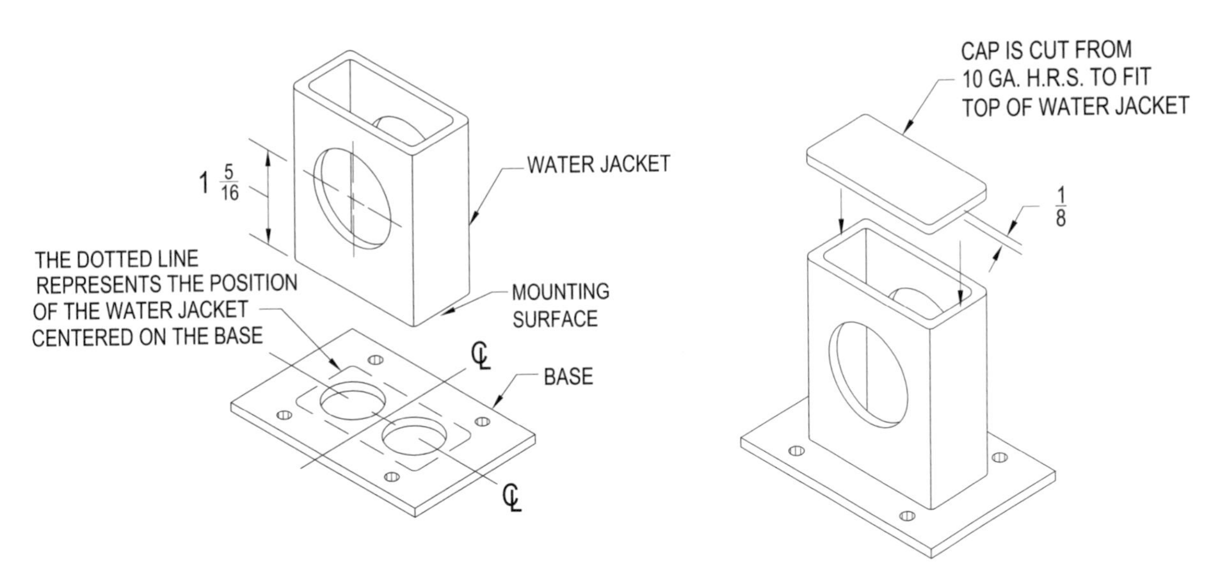

Figure 97 *Center the water jacket on the base and braze in position.*

Figure 99 *Cut a piece of 10 gauge h.r.s. to fit top of water jacket then braze it in position.*

Since the cylinder will be brazed or silver soldered in the water jacket bore it must be a close but easy sliding fit. A forced fit will not braze or solder well. The bore must be parallel to the mounting surface and perpendicular to the vertical axis. The cylinder is bored to near size and finished with a lap or hone. Since the intense heat of brazing would distort the finished cylinder, assembly must be complete before final boring and honing. The outside diameter of the cylinder is not critical, but it should be at least filed smooth in the lathe and the spark plug end should be turned smooth to ride in the steady rest. While the cylinder might be held only in the chuck it is likely to bore tapered so a steady rest should be used.

Figure 98 *Using a bolt & nut to secure the water jacket in position for brazing.*

The intake, exhaust and spark plug bosses are prepared and brazed in place before boring the cylinder. But their corresponding holes are milled through the cylinder wall after boring, but before honing or lapping.

The valve and spark plug bosses are shown in figures 100, 101 and 102. To form the concave on each one to match the cylinder curve, a fly cutter can be set to cut a circle equal to the outside diameter of the cylinder. If you mount a length of stock vertically in a mill vise or the milling attachment on a lathe, the fly cutter will cut a concave in one side of stock. Several passes will be required to cut

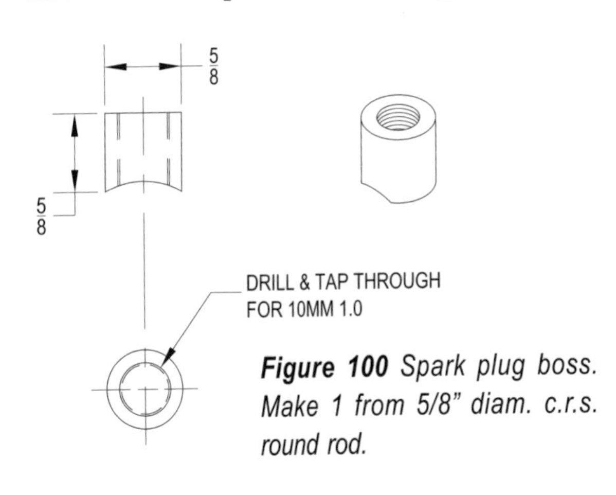

Figure 100 *Spark plug boss. Make 1 from 5/8" diam. c.r.s. round rod.*

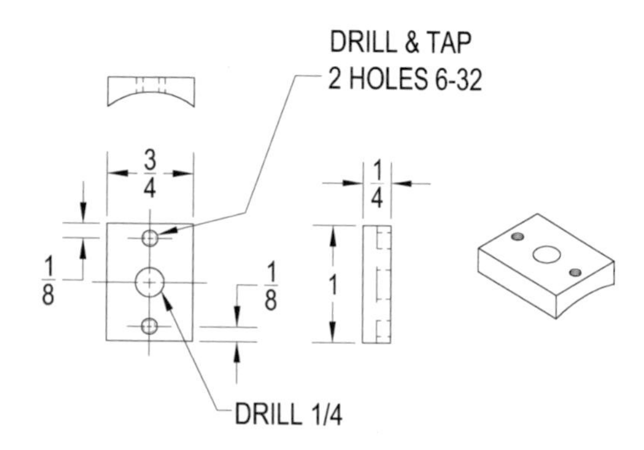

Figure 101 *Exhaust valve boss.*
Make 1 from 1/4" x 3/4" c.r.s. bar

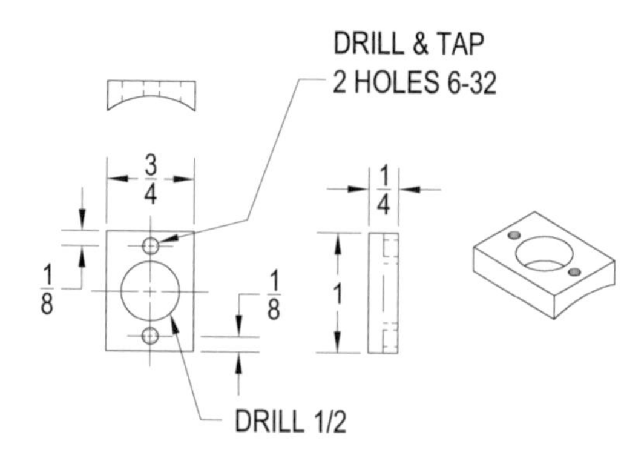

Figure 102 *Intake valve boss.*
Make 1 from 1/4" x 3/4" c.r.s. bar

the concave to the full width of the stock and a bit over 1" long. To make the spark plug boss shown in figure 100, the end of a 5/8" diameter round is similarly prepared and parted off. Then chuck it in the lathe to drill and tap for the 10mm spark plug. Cut the valve bosses off at 1" length and drill and tap them as indicated in the figures.

The cylinder is placed in the water jacket bore and brazed in position and the and the valve and spark plug bosses are brazed to the cylinder. See figure 103 for the cylinder postion in the water jacket and figure 104 for positioning of the valve and spark plug bosses.

Bore the cylinder to within a few thousandths of final size on the lathe using the chuck and steady rest, or you could use the mill. The work must be carefully centered especially if 1" pipe is used because it will barely clean up at .120". If cold rolled solid round or cast iron is used for the cylinder it must be bored to near size before assembly or it will be difficult if not impossible to braze it to the water jacket.

When the assembly is bored clean and within .002" of nominal size the spark plug, intake and exhaust ports are milled through the cylinder wall.

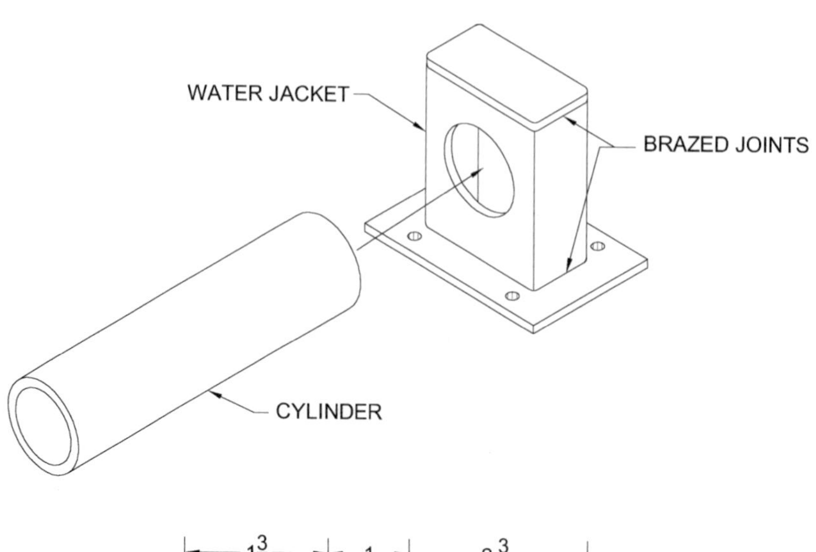

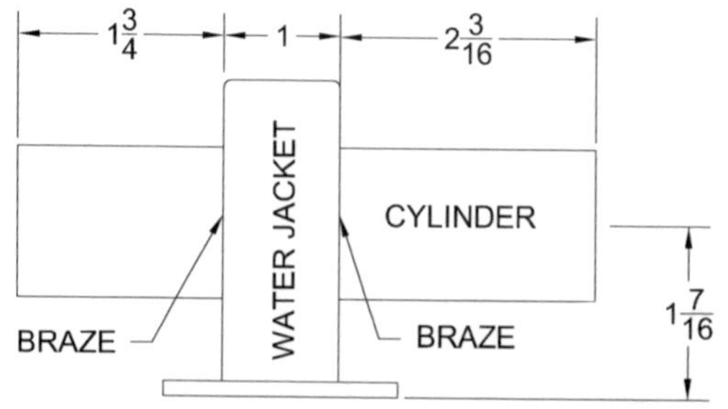

Figure 103 *Position the cylinder in the water jacket as shown, then braze each joint..*

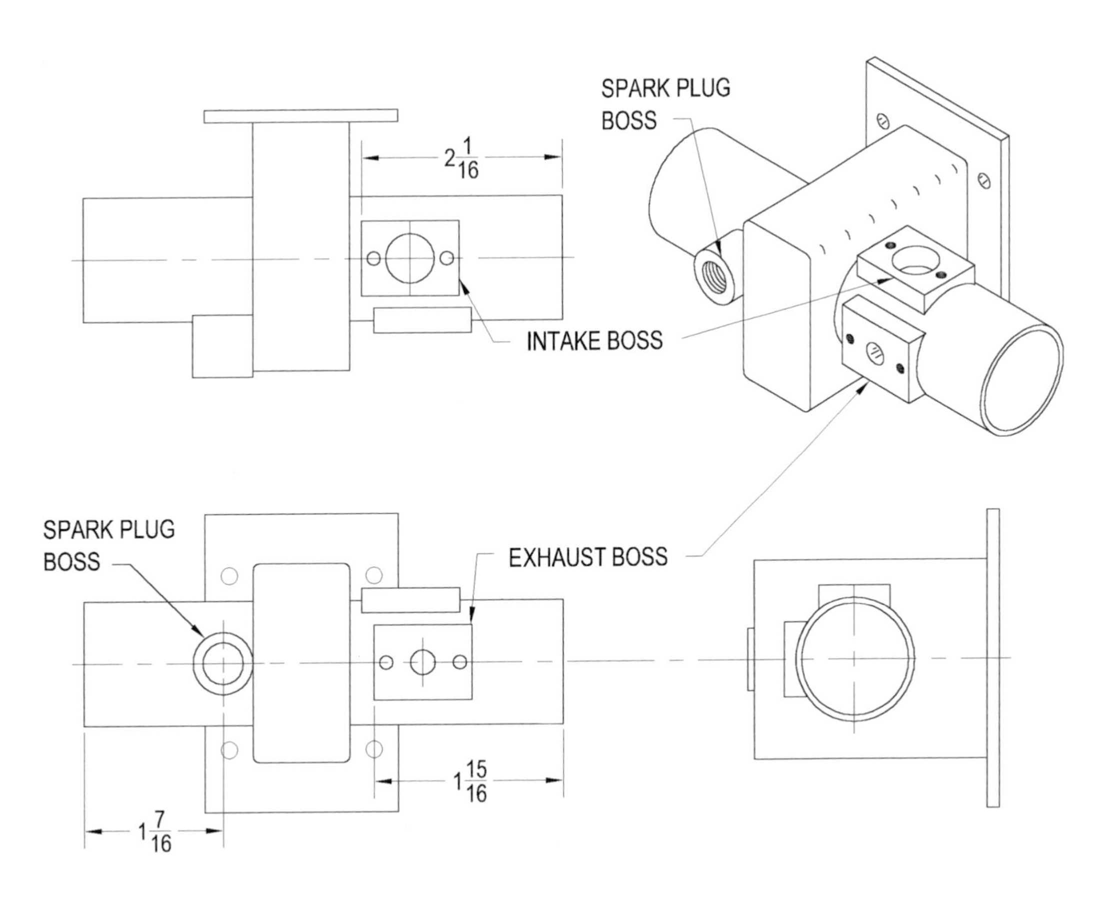

Figure 104 *Cylinder layout showing the position of the intake, exhaust and spark plug bosses.*

Figure 105 *Boring the cylinder.*

Figure 106 *The cylinder assembly as it looks so far. The intake, exhaust and spark plug bosses have been brazed on and the cylinder has been bored to near size.*

Before milling or drilling the spark plug hole, it's a good idea to prepare a short alignment bushing turned to the tap size of the 10 mm spark plug. Then slip the bushing into the spark plug boss positioning the 3/16" offset hole closest to the water jacket side of the boss. See figure 107 for hole location. Guide a 3/16" drill or end mill into the hole in the alignment bushing and through the cylinder wall.

You can make an alignment bushing for the intake port too, but first mill a slight flat for valve clearance on the O.D. of the cylinder wall inside the intake port. When milling the flat, be careful not to mill completely through the cylinder wall. Once the flat has been milled, position the alignment bushing in the intake hole. Locate the bushing to offset the hole to the side closet to the water jacket. Then mill or drill the 3/16" intake port through the cylinder wall.

The exhaust port is milled or drilled through in the same manner using an alignment bushing. Insert an alignment bushing in the 1/4" hole of the exhaust boss positioned so the 3/16" hole is offset to the left. (The side closest to the water jacket.) Then guide the mill or drill through the cylinder wall.

When all three ports are drilled, use a "3" cornered scraper or other de-burring tool to carefully scrape any sharp burrs that may be around the holes inside the cylinder bore. Take care not to score the cylinder bore. All you want to do is remove excessive burrs.

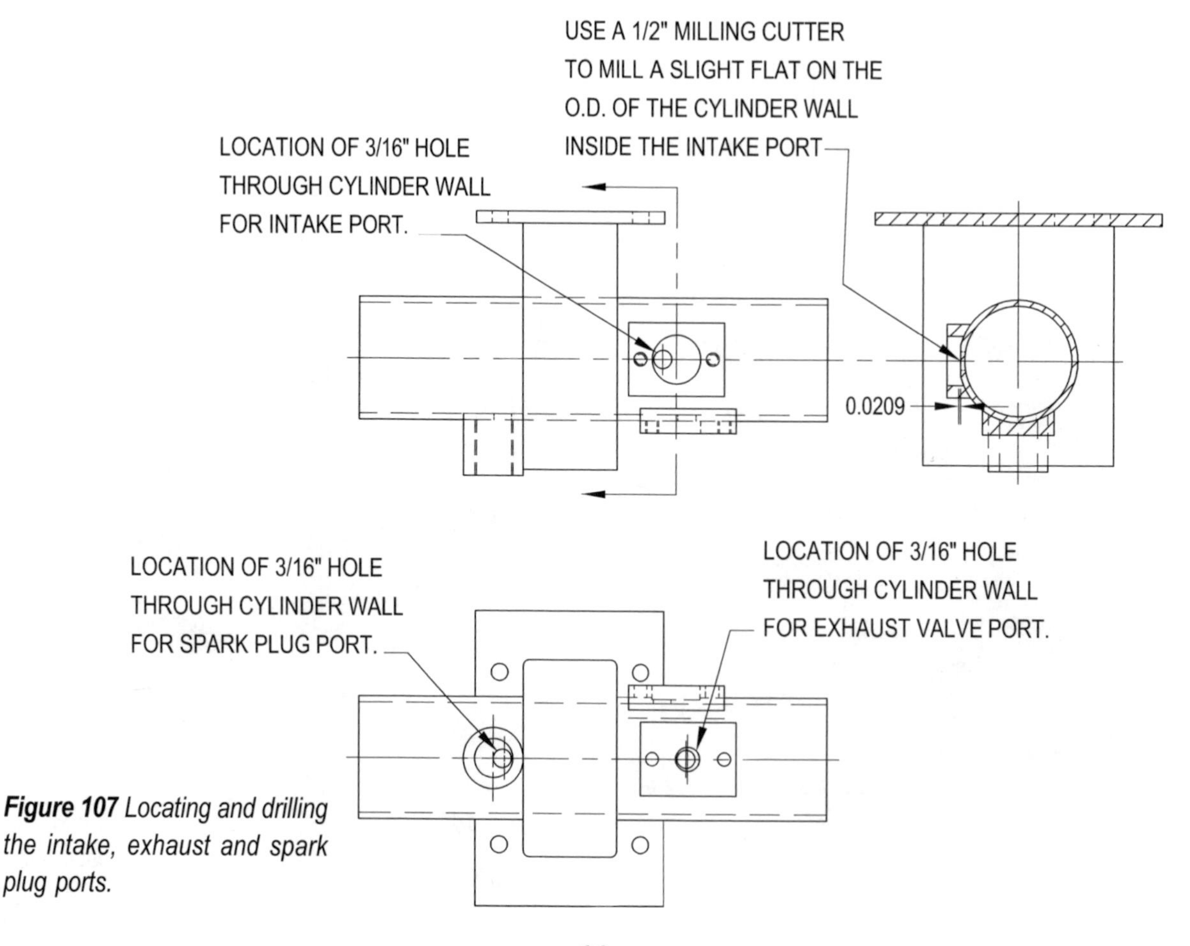

Figure 107 *Locating and drilling the intake, exhaust and spark plug ports.*

Finishing the cylinder . . .

An important objective in any engine project is an accurate cylinder bore. Even with the most careful set up, a lathe or milling machine will not bore a perfectly round hole and chances are there will be significant taper. To finish the cylinder wall and straighten the bore, we need either a cylinder hone or lap. There is a distinct difference between lapping a cylinder and honing a cylinder.

A cylinder hone uses abrasive pads for cutting. It is often referred to as a pin hone. Such a hone cuts very fast and excellent results are achieved. A cylinder hone can actually remove a taper of several thousands of an inch in a matter of minutes. This type of hone is very expensive and most will not have one in the home shop. But they can be found in motorcycle, automotive and in some machine shops. Often the shop will be glad to hone your cylinder for a small fee. I did some checking and this fee seems to range from $25.00 to $35.00. Not too bad considering the time it saves and accurate results that can be achieved. If you end up with a significantly tapered bore this may be your best option short of making a new cylinder.

Rather than a cutting stone, a cylinder lap uses an abrasive charge. In order for a lap to work, it must be made of a material softer than the piece you're working on. In this case softer than the cylinder walls. Because the lap is softer than the cylinder walls, the abrasive charge will imbed in the lap and not in the walls of the cylinder. The imbedded abrasive in the lap is what does the cutting. Don't expect to remove a significant taper or deep gouges with a lap. If your cylinder has a taper of over .003" or .004" and has serious flaws that extend the length of the cylinder then you may need to consider remaking the cylinder or taking it to a shop for honing.

We experimented quite a bit with cylinder laps and came up with several ideas. One thing I learned for sure is that an entire book could be written on the subject. In the end, the simplest lap to make and one that did the best job was a lap design used by our friends Jim Lewis and Ben Imbrock. It's a copper lap and they seem well satisfied with its performance. And after trying it I have to agree. Essentially it is a short length of copper pipe with a slit along its length. One edge of the slit is brazed to an expanding mandrel. Tightening a set screw in the lap mandrel against a wedge expands the copper pipe to fit the cylinder bore. See photo in figure 108.

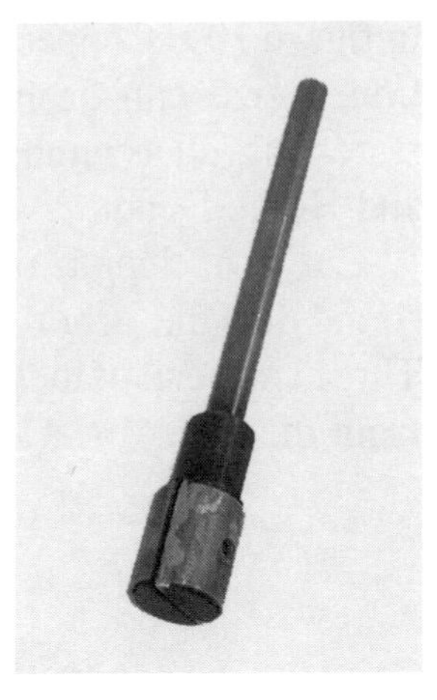

Figure 108 *Copper lap.*

One thing to mention before getting started is that even though we need to strive for a 1.125" cylinder bore, don't be too concerned if your cylinder finishes out somewhat over sized. It should still work OK. My cylinder finished out at 1.130" when complete, and at least one of our other engines was oversized by more than that. The main thing to be concerned with are cylinder walls that are clean, parallel and measuring the same at both ends. (No taper)

Making a copper lap . . .

Parts to make the lap are as follows:

A 1-1/2" length of 1" copper pipe. 1" copper is perfect because it's O.D. measures almost exactly 1-1/8" which is the I.D. of our cylinder.

A 2-1/2" length of 1-1/8" diameter c.r.s.round. This will become the expanding mandrel.

A 6" length of 1/2" diameter c.r.s. round. This is the arbor.

A 1/4-20 x 3/4" socket head set screw.

Begin by cutting the 1-1/2" length of 1" copper pipe. Drill a 1/4" hole in the side of the pipe, centered along its length as shown in figure 109. Copper pipe in the above mentioned size can be purchased at hardware stores as well as lumber stores such as Lowes and Home Depot.

Cut a 6" length of 1/2" diameter c.r.s. round for the arbor shown in figure 110. Thread one end of the arbor 1/2-20, 3/4" back from the end.

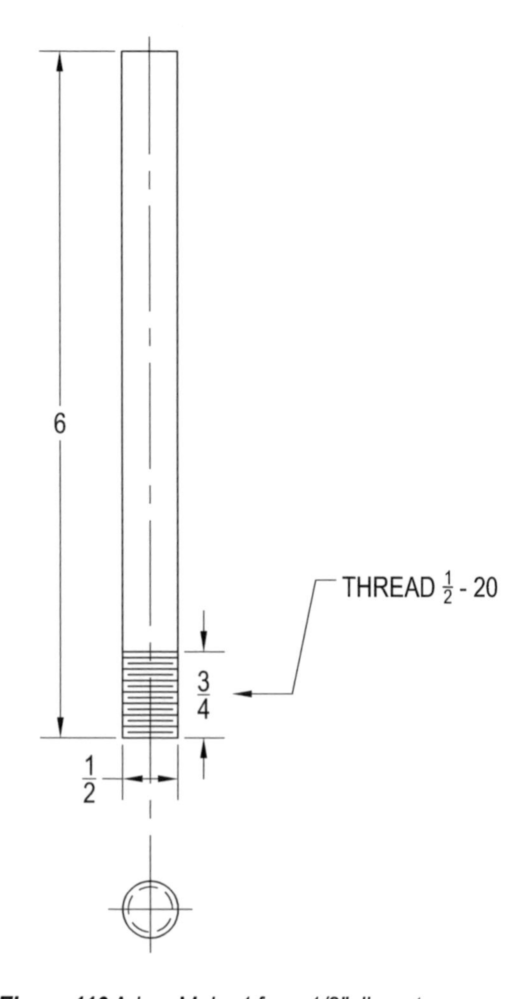

Figure 110 *Arbor. Make 1 from 1/2" diameter c.r.s.*

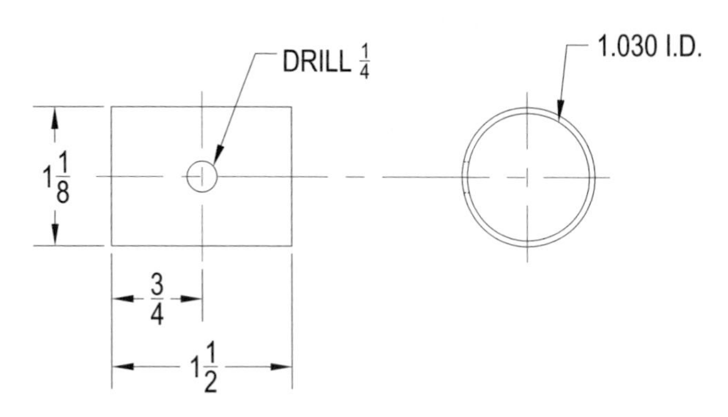

Figure 109 *Lap shoe. 1-1/2" length of 1" copper pipe.*

Cut a length of 1-1/8" diameter c.r.s. round rod slightly longer than 2-1/2" for the expanding mandrel shown in figure 111. Face off both ends to a final length of 2-1/2". Reduce the diameter of one end to 7/8", 1" back from the end, then center drill that end and drill 29/64", 3/4" deep. Tap the 29/64" hole for 1/2-20 thread. Remove the mandrel from the lathe, then drill and tap the 1/4-20 hole in the side of the mandrel as shown in the figure.

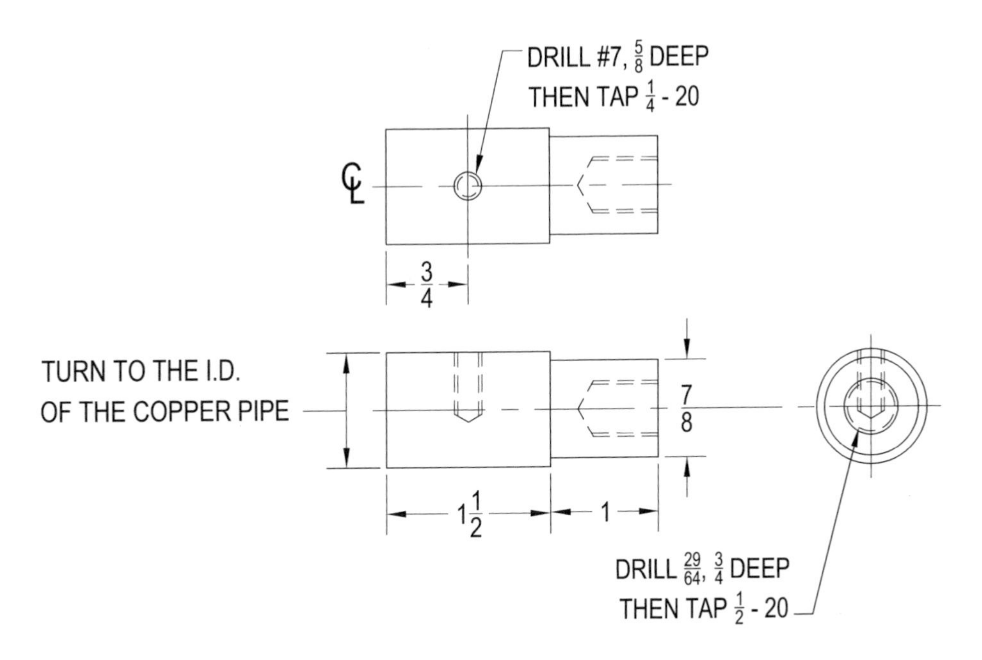

Figure 111 *Expanding mandrel. Make 1 from 1-1/8" diam. c.r.s. round.*

Thread the arbor into the 1/2-20 tapped hole in the end of the mandrel. Mount the arbor in the lathe and reduce the diameter of the large end of the mandrel to match the I.D. of the copper pipe.

To convert the mandrel into one that will expand, cut away the wedge section as shown in figure 112. Set it to one side. Run the 1/4-20 tap through the threaded hole in the stationary side of the mandrel again to cut the threads on through. There should be a slight dimple in the wedge section for the end of the set screw to rest in.

Slide the copper pipe section (lap shoe) onto the stationary portion of the mandrel. Align the 1/4" hole in the shoe with the 1/4-20 hole in the lap. Scribe a line along the side of the lap shoe to correspond with the edge of the stationary mandrel. Cut out a 1/8" wide slit along the length of the shoe as shown in figure 113.

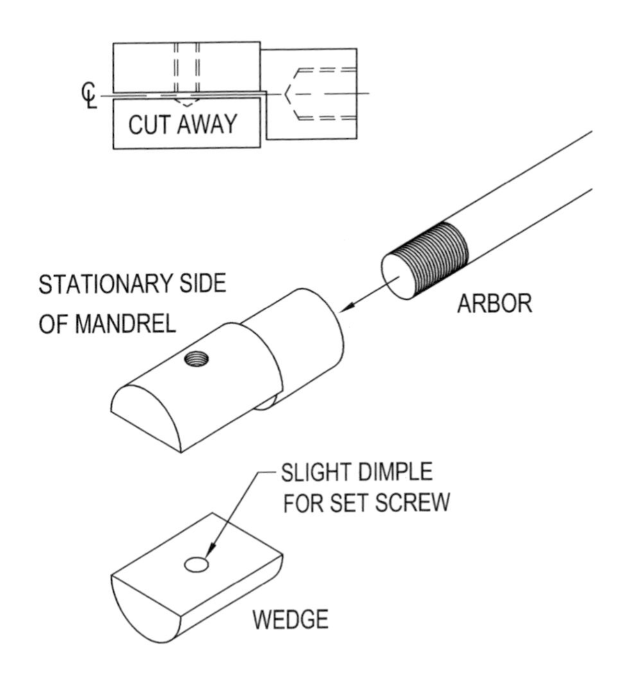

Figure 112 *Cutting out the wedge section of the mandrel.*

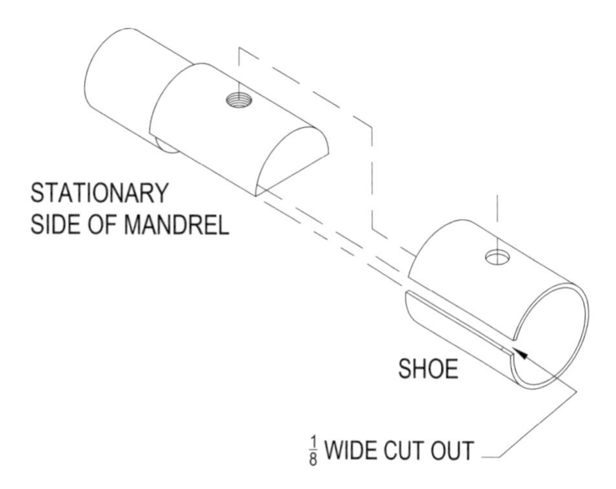

Figure 113

Reposition the shoe on the mandrel aligning the 1/4" hole in the shoe with the 1/4" hole in the mandrel. Secure with a 1/4" bolt. Braze the edge of the copper pipe to the mandrel as shown in figure 114.

After the lap cools, remove the 1/4" bolt and insert a 1/4-20 x 1/2" set screw in the mandrel. Insert the wedge you set aside earlier inside the shoe. The lap is expanded by tightening the set screw against the wedge. If you drilled the # 7 hole deep enough earlier, there will be a slight dimple in the flat surface of the wedge. The end of the set screw will rest in the dimple and prevent the wedge from falling out.

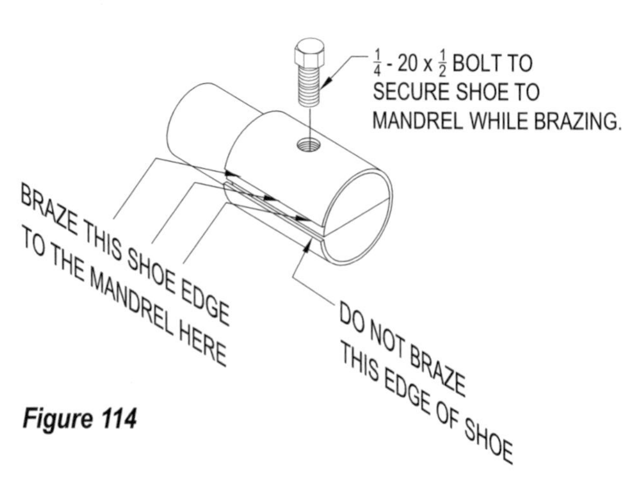

Figure 114

Before using the lap it must be charged with an abrasive compound. A grade 220 or 280 valve grinding compound will work fine. The charging process is accomplished by spreading a small amount of compound on a flat surface that is harder than the lap itself. A slab of steel will work fine. Roll the lap through the abrasive compound as if you were rolling pie crust. The abrasive particles will embed in the softer copper lap.

To begin the actual lapping process, chuck the lap arbor in the drill press. Set the drill at its slowest speed. You will find the lap has a tendency to grab at the cylinder from time to time so the slow speed will help keep things under control. Keep in mind, there are edges on the cylinder assembly that could cause injury if the lap were to hang up and twist out of your hands. Wear gloves and be prepared to turn loose.

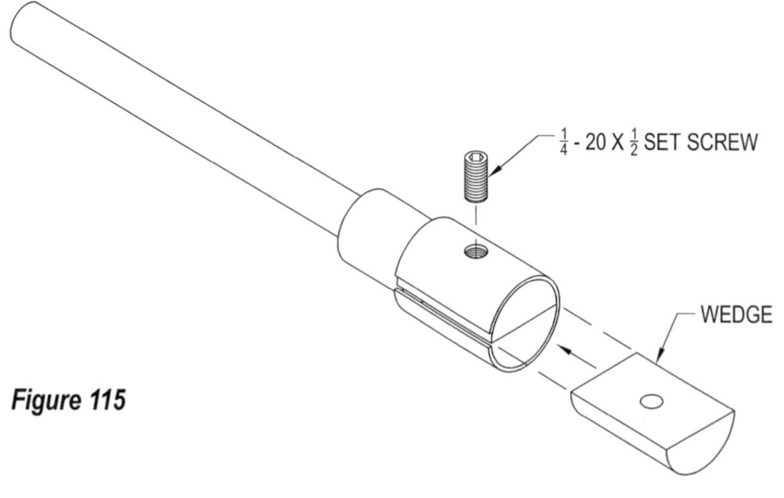

Figure 115

You're almost ready to start. Adjust the lap to fit the cylinder bore. It should fit snug, but you should still be

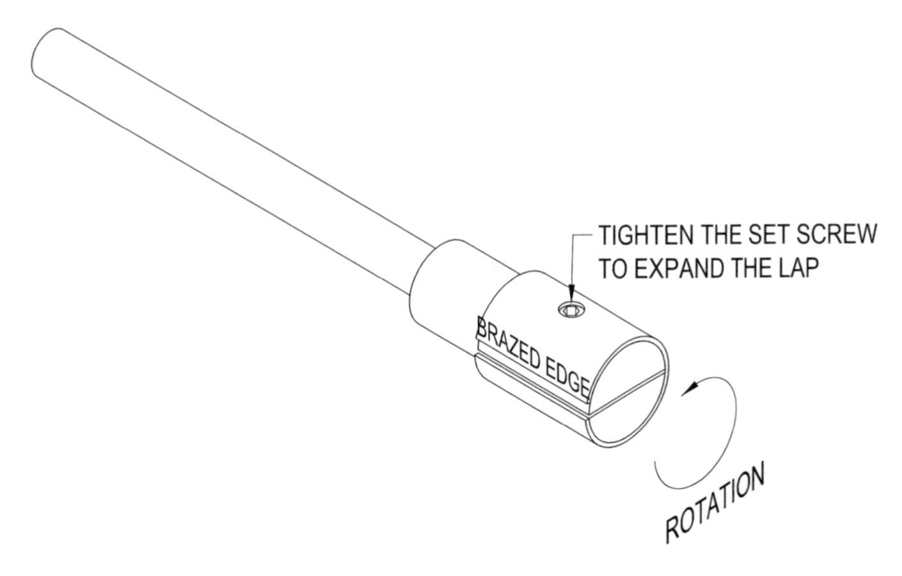

Figure 116

able to rotate it inside the cylinder. If the lap fits too tight it will grab the cylinder out of your hand when you turn on the drill. And be careful too, because if you adjust the lap to fit too tight it may be very difficult to remove it from the cylinder. Then again if it's too loose it will not cut well or evenly. After a couple of tries you will get the hang of it. You will need to flush the cylinder while lapping, and kerosene works well for this. Pour a little in a coffee can and use a brush to drip it in the cylinder. When you are ready, take a firm grip on the cylinder and turn on the drill. The lap will begin rotating inside the cylinder. It is very important to keep the cylinder moving in an up and down motion continually while lapping otherwise uneven grooves will be cut in the cylinder wall. Remember, smooth, straight and parallel cylinder walls are the objective. While lapping, don't let the lap tool extend more than 1/8" beyond either end of the cylinder or the cylinder could end up wider at the ends of the bore. As the lapping tool revolves, flush the work by dripping kerosene through the cylinder with a small paint brush from time to time as needed.

When you are lapping you will feel more resistance in some areas than others. The objective is to continue lapping until these high areas are cut away. You can expand the copper shoe from time to time by tightening the set screw. And the lap will need recharging from time to time as it looses its abrasive feel and cutting ability. Continue lapping until a visual inspection shows a uniform texture from end to end. The object is to produce a cylinder that is truly round with parallel walls. Your final finish will depend on the condition of the bore when you begin. Check both ends of the cylinder bore to ensure there is no taper. Lapping is a slow process and it can take an hour or more to increase the bore of a cylinder a few thousandths of an inch. Tiny flaws or a slight ripple will not seriously hamper the operation of the engine. But any flaw that extends the length of the cylinder will cause problems. When the work is finished flush the cylinder thoroughly to remove all traces of abrasive.

The piston rings . . .

1-1/8" piston rings for this project can be purchased inexpensively from, "Otto Gas Engine Works", 2167 Blue Ball Rd., Elkton, Maryland 21921-3330. Phone 410-398-7340. View their catalog on the web at http://www.dol.net/~dave.reed/otto.html

But some may have the desire to make their own. The following discussion covers both heat treated and non-heat treated rings. In my experience, the heated treated rings seem to have more spring and may be more reliable for gas engine application than those that are not. But both methods are included for those who might want to experiment.

There are subtle details in modern piston ring engineering, such as elliptical forms or thinned section opposite the gap, not included in these basic procedures, it is practical to produce good quality rings in the home shop. Automotive brake cylinders are one source for fine grain easy machining cast iron for rings or you may purchase cast iron from one of the sources listed at the back of the book.

Of the several ways to produce piston rings, the most common way is to calculate the correct outside diameter to allow for closure of the gap, deduct twice the thickness of the ring to arrive at the inside diameter and then machine the blank to those dimensions. The rings are then parted off, slit and expanded by heating to a red heat.

If your slitting saw is .025" you would add that amount to the calculated circumference.

If 3.1416 x diameter= circumference.
Ring thickness is .050
Ring diameter is 1.125 so
1.125 x 3.146 = 3.5343 and
3.5343 + .025 = 3.593 then
3.593 divided by 3.1416 = 1.133 and
1.133 - .100 = 1.033

Thus the blank will be machined to an outside diameter of 1.133 and an inside diameter of 1.033. Such a ring will fit a bore of 1.125" when compressed to close the gap. But it will not exert enough pressure on the cylinder wall for effective sealing. If the gap is spread to 3/16" by inserting a tool bit and the ring is heated uniformly to a bright red the ring will take a "set" to the expanded diameter and work well after break in.

To ensure uniform heating and avoid distortion you can clamp the expanded ring to a metal or asbestos plate, leaving a gap between the ring and the plate. Center the propane torch flame in the ring and strive for uniform color as the ring begins to turn red. Allow to air cool.

The main objection to this basic method is the possibility of distortion in the heating process that can prolong "wearing in" or be so drastic as to make the rings useless. There is a method for "truing" distorted rings. And the same method can be used to produce rings without heat treating. This involves producing slightly over sized rings and clamping them in a fixture to be machined in a compressed state to final size. While the fixture must be made on the lathe in which it is to be used, it can be used many times if its original orientation in the chuck is repeated, If the fixture is used for heat treated rings they are just made slightly over size to allow for finishing. Otherwise the rings are made much larger and the gap is cut wider to allow for compression.

If a non-heat treated ring is desired we simply add the amount of the expanded gap to the circumference of the nominal O.D.

3.1416 x diameter = circumference.
Diameter = 1.125
expanded gap = .1875
thickness = .050 x 2 = .100
1.125 x 3.1416 = 3.5343
3.5343 + .1875 = 3.7218
3.7218 divided by 3.1416 = 1.1847

Add .010" to the basic diameter for finish machining for a blank outside diameter of 1.1947". Subtract .100 from the basic diameter for a blank I.D. of 1.0847". Thus the rings will fit the mandrel of the fixture when compressed and the clamp will hold them in a compressed state while machining to true size.

Using a sharp tool with moderately fine feed will finish the rings to an appearance of a very fine thread. Do not file to exact size. The ridges of the very fine thread will quickly wear to conform to the cylinder wall and break in will be more rapid.

Part off rings to a width slightly wider than that desired in the finished product. Face off the new end before parting off each of the remaining rings.

When the rings have been parted off we can use a ring collet fixture to hold each one in position while we face off the unfinished side and bring each ring down to its proper width. The collet as shown in figure 117 is easy to make and with such a fixture, excellent results can be achieved and the width of all your rings will be the same.

The idea for this fixture comes from an article by Garold B. Kubicek published in Live Steam Magazine titled, "Machining Steam Engine Piston Rings".

The collet can be made from a variety of metals, but I chose aluminum round. For proper use, it must be indexed to the lathe chuck it will be used in. The drawing in the figure gives the dimensions for the collet.

To use the fixture, place it in the lathe chuck with the index mark aligned with the corresponding mark on the chuck. Insert the ring in the fixture and ensure it is well seated by tapping it in with a mallet. Tighten the chuck to secure the ring, then face to width.

The photos and drawings on this page and the next will take you through the process of cutting the ring gap and making the ring turning arbor for finishing the rings to their final diameter.

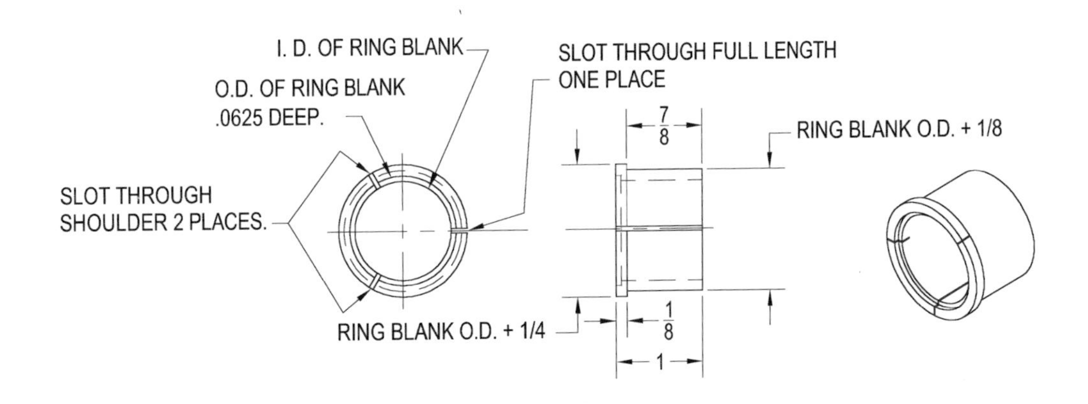

Figure 117 *Collet fixture for facing rings. Mine was made from aluminum round, but other metals would work as well.*

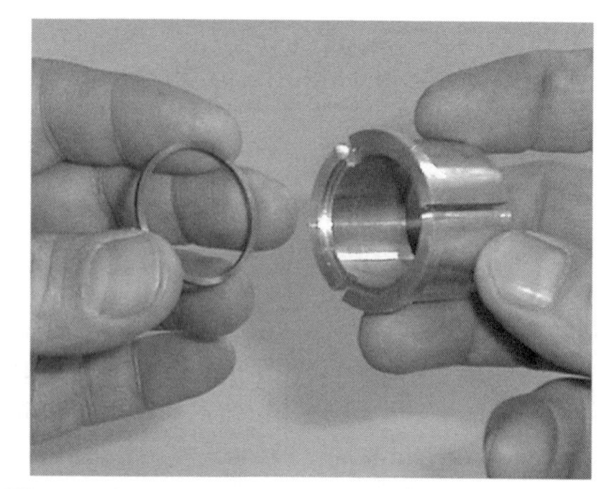

Figure 118 *Piston ring and collet fixture.*

Figure 120 *Checking ring gap with a 3/16" drill.*

Figure 119 *Holding the ring in a vise and cutting the gap using a multi purpose tool with slitting saw. Note how the ring is held between a couple of pieces of wood to protect it from being damaged by the vise jaws. Cut the gap a bit narrow and then finish to final distance with a file.*

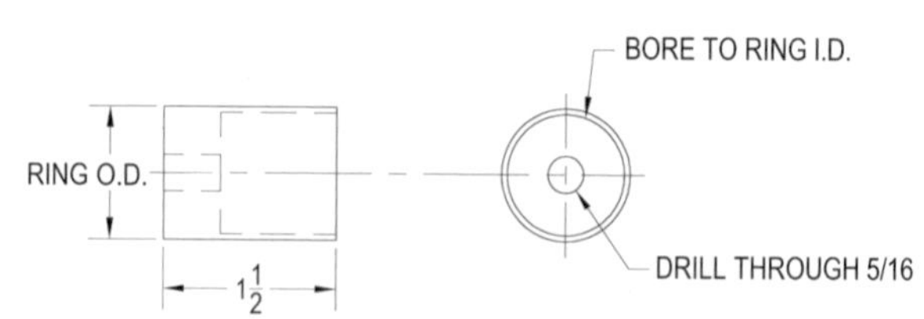

Figure 121 *Clamping end of ring turning fixture.*

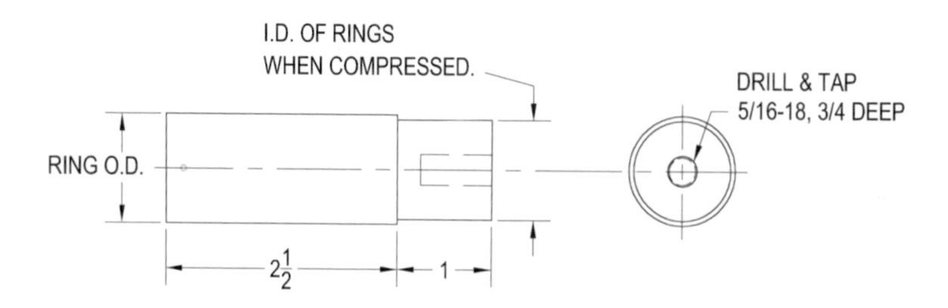

Figure 122 *Mandrel side of ring turning fixture.*

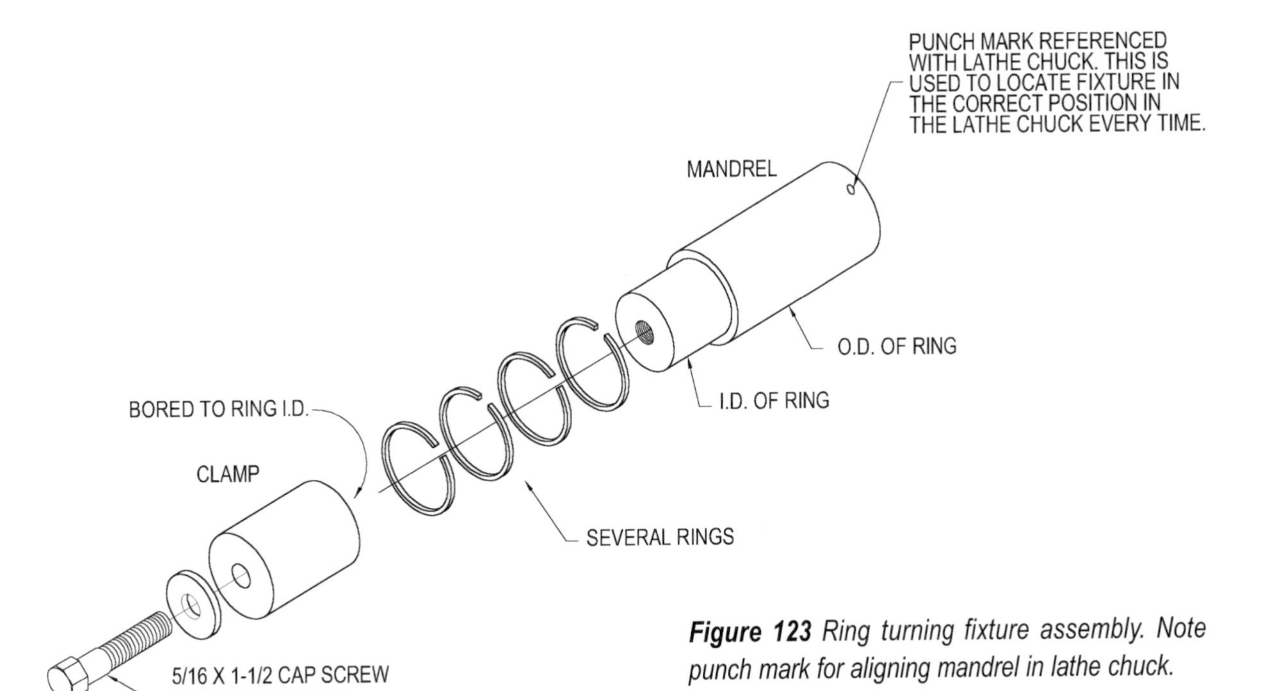

Figure 123 *Ring turning fixture assembly. Note punch mark for aligning mandrel in lathe chuck.*

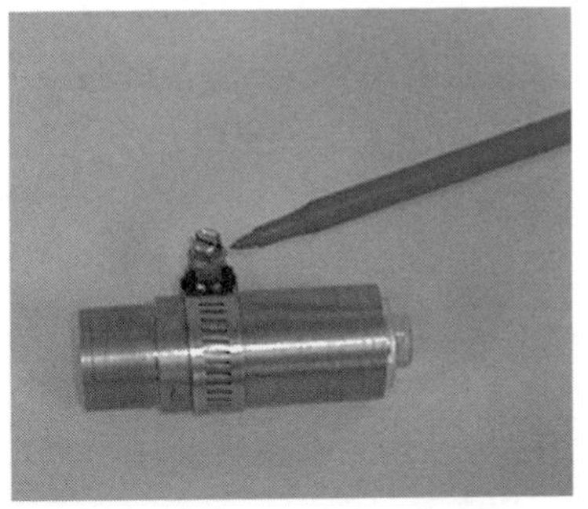

Figure 124 *Rings can be compressed around mandrel with a hose clamp.*

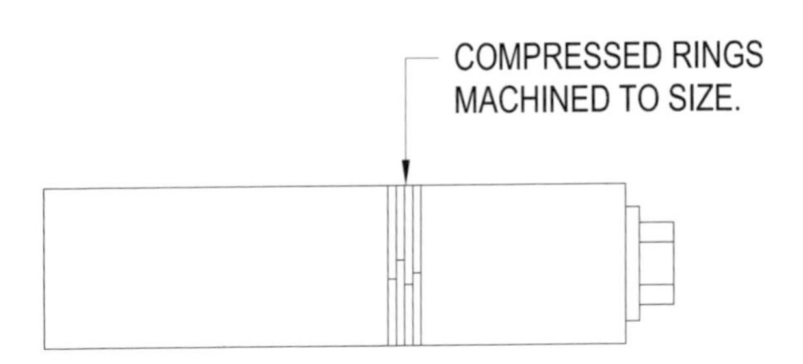

Figure 126

Figure 125 *Tighten cap screw to clamp rings in position. Remove the hose clamp, clamp fixture in lathe chuck and turn the rings to size.*

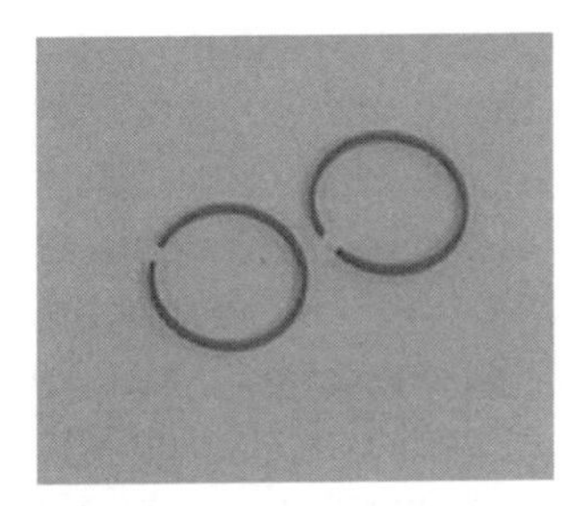

Figure 127 *A couple of finished rings.*

Making the pistons . . .

The pistons are made from 1-1/4" diameter aluminum round. See figure 128. Before making the piston it is a good idea to have the rings in hand so that you know what size to make the ring grooves.

Begin making the piston by chucking a length of 1-1/4" diameter aluminum round in the 3-jaw chuck so that 1-3/4" sticks out past the jaws of the chuck. Face off the end and turn the outside diameter of the piston to a diameter of .002" less than the diameter of the cylinder. The object is to have the piston sized so that it travels in the cylinder freely with minimum side play.

Cut the grooves for the piston rings using special care. The finished ring grooves should be about .001" wider than the piston ring and about .008" deeper than the thickness of the rings. The cut off tool to form the ring grooves should be slightly narrower than the ring and two or more passes are needed to cut the groove to finished width. The set up for cutting the ring grooves is the same as that for a normal cut off operation. The tool should be set at a right angle to the work in the proper position to cut the groove. Lock the carriage and cut the groove to a depth .008" greater that the thickness of the ring. Check the ring for fit as you proceed. The ring should fit the finished groove freely with minimum side play.

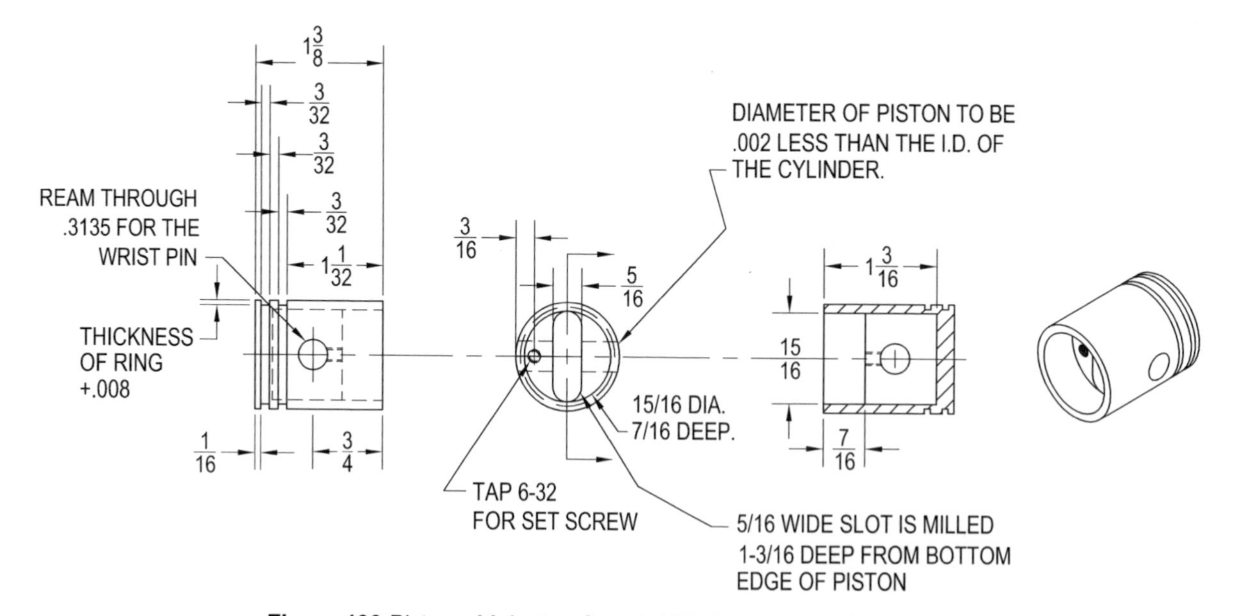

Figure 128 *Pistons. Make two from 1-1/4" diameter aluminum round.*

After the ring grooves are cut, step drill into the end of the piston to 5/16", 1-3/16" deep. Then counter bore to 15/16" diameter 7/16" deep. Part off the piston 1-3/8" long.

To cut the slot in the piston, clamp it in a mill vise using a V- block to hold it securely. Center the piston with the spindle of the mill then cut the 5/16" wide x 15/16" long slot 1-3/16" deep using a 5/16" end mill.

Align the piston in a V-block and clamp it in position to drill and ream the .3135" wrist pin hole.

And finally drill and tap the 6-32 hole for the set screw that will be used to secure the rod pin.

The piston rods . . .

A dimensional drawing of a piston rod is shown in figure 130. Two are required and each can be milled from the same casting, getting one from each end. Or you could machine them from solid stock. To better visualize how two are obtained from the same casting, look at figure 129 which contains a drawing of the casting with an outline of a piston rod at each end. Instruction for machining the piston rods follow:

1. Mill casting, top/bottom, sides and ends to square.
2. Layout position of .3135" holes on each end of casting.
3. Layout position of milled slot in each end of casting.
4. Drill near size, then ream the .3135" holes. When drilling, the casting should be clamped in a vise that is secured to the mill or drill table. Ensure the drill is perpendicular to the casting. Straight holes here are important.

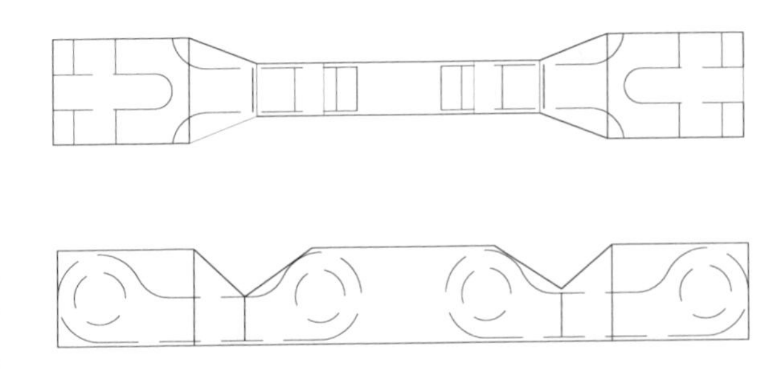

Figure 129 The piston rod casting showing the outline of a piston rod at each end.

5. Reposition the casting in the vice to mill the 1/4" wide slot, 3/4" deep. See figure 131.
6. Turn the outside radius on each side of the casting as shown in figure 135. See figure 132-134 for help in making the fixture for turning the radius.
7. Cut off each piston rod to length. Use fixture and a handle arrangement shown in figure 137 to turn end radius.
8. Finally, mill the side profile to the correct width. A completed piston rod is shown in figure 138.

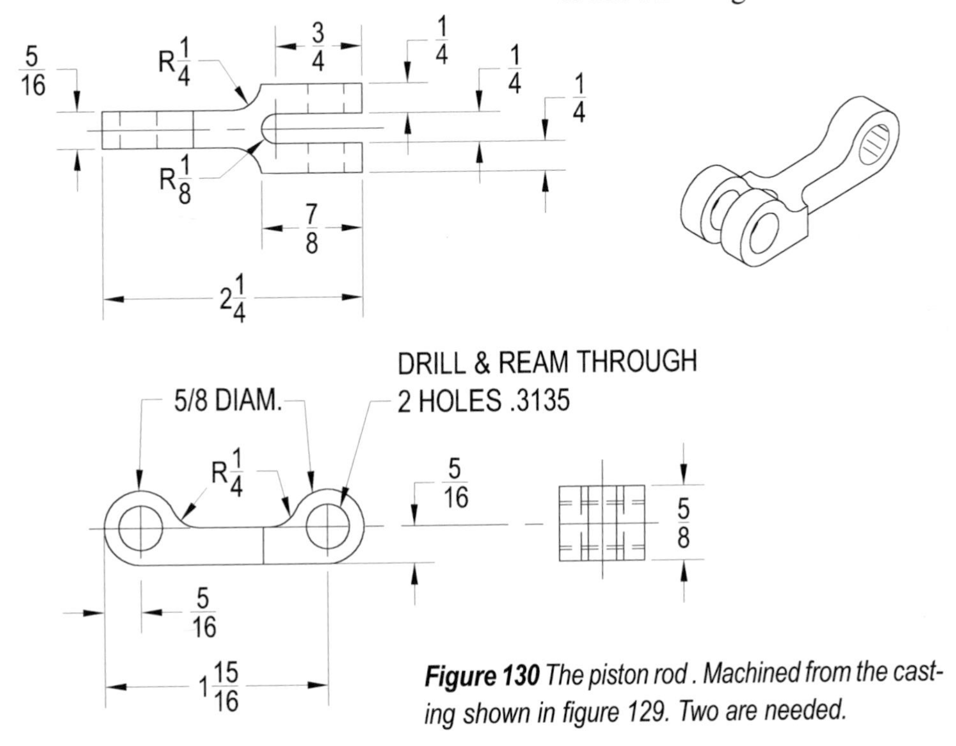

Figure 130 The piston rod . Machined from the casting shown in figure 129. Two are needed.

Figure 131 *Piston rod casting is held in the vise and we are milling a slot in a piston rod. The .3135" holes have already been drilled and reamed.*

A radius turning fixture . . .

The radius turning fixture is a simple solution for turning a radius on the ends of the piston rods. It consists of two main parts, the stationary arm and the pivot shaft. In operation, the stationary arm of the fixture is held in the mill vise. The radius is formed by rotating the piston rod casting on the pivot shaft and forcing it against the rotation of the milling cutter. The #10-24 socket head cap screw and washer shown in figure 134 and in the photo in figure 135 is used to prevent the casting from working its way off the pivot. Half of the radius is formed first and then the casting is turned over to form the other half. Begin by centering the end of the casting against the center of the milling cutter. Then back the work away from the cutter until you can pivot the casting without hitting the cutter. Turn the machine on. Begin by advancing the work slowly towards the milling cutter a few thousands at a time. Stop the feed and carefully pivot the casting against the milling cutter. Continue advancing the work a few thousands at a time for more passes until the half radius is formed.

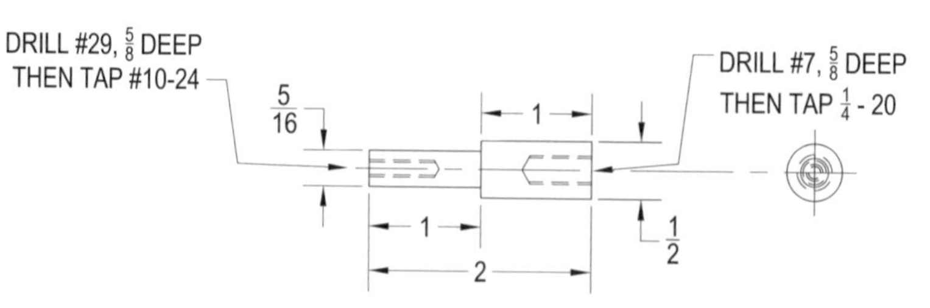

Figure 133 *The pivot shaft for the radius turning fixture. Make one from 1/2" diameter c.r.s. round, 2" long.*

1/2
1
DRILL THROUGH 31/64
THEN REAM .4990
5/8
4

Figure 132 *The stationary arm for the radius turning fixture. Make one from 5/8" x 1" x 4" h.r.s. flat bar.*

Then turn the casting over and repeat the process to form the other half of the radius. Form both ends of the casting in the same manner. Remember, take light cuts and feed the work slowly. There is a risk of the milling cutter grabbing the work so be extra cautious. Don't let it pull your hands into the cutter. Be prepared to turn loose and quickly turn the machine off. And don't attempt to rescue work while the machine is running. After the piston rods have been separated from the casting add a handle to each as shown in figure 137 and use the same method to turn the remaining end radius on each rod.

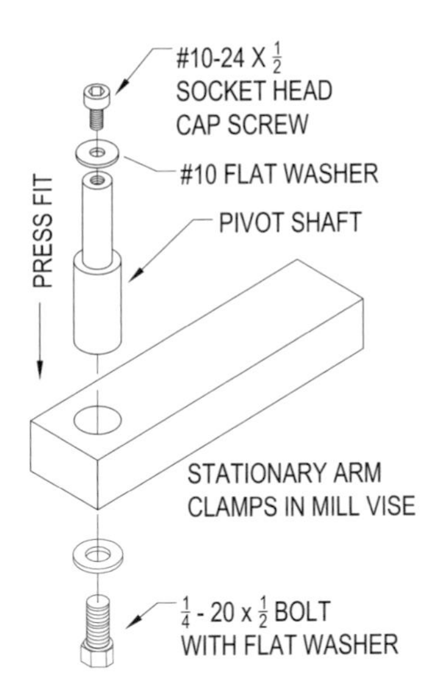

Figure 134 *Assembly drawing of the radius turning fixture complete.*

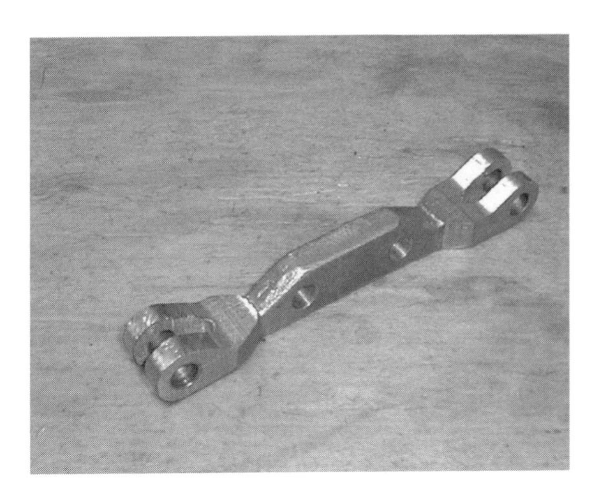

Figure 136 *Piston rods are ready to be cut from each end of the casting.*

Figure 135 *The slots have been milled and we are now using a pivot fixture held in the mill vise to form the radius on the ends. Notice how the #10-24 socket head cap screw and washer are used to prevent the casting from working its way off the pivot. Use a slow feed and maintain a firm hold. It's very important to only apply the work against the rotation of the milling cutter. Be aware, there is still a risk of the milling cutter grabbing the work. Don't let it pull your hands into the cutter. Be prepared to turn loose and quickly turn the machine off. Don't attempt to rescue work while the machine is running.*

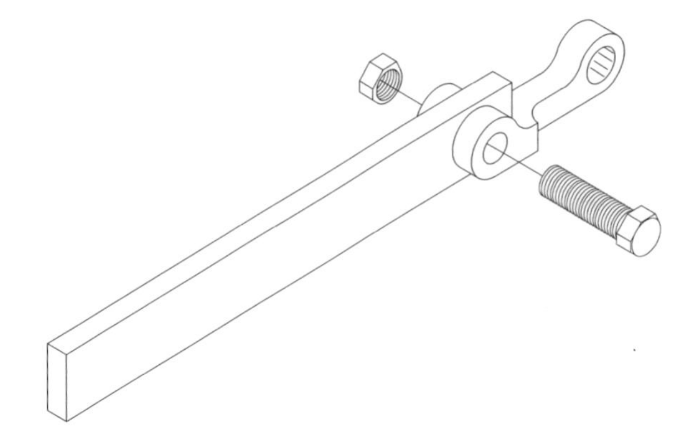

Figure 137 *After the piston rods are cut from the casting, you can make a handle for turning the other end radius from a piece of 1/4" x 3/4" flat bar. Bolt it in the slot as shown in the drawing.*

Figure 138 *A completed piston rod.*

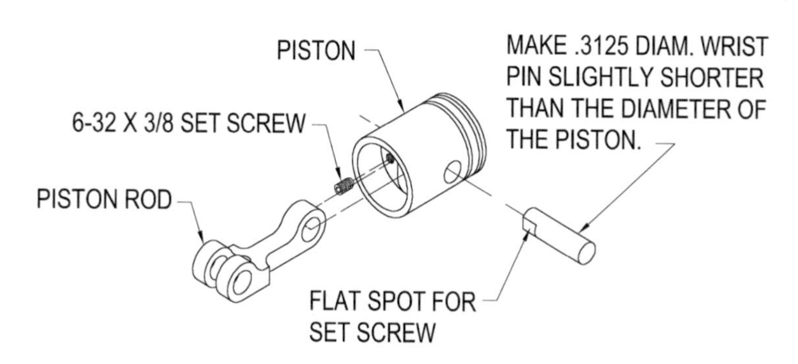

Figure 139 *Assemble the piston rod to the piston. Wrist pin is made from 5/16" diameter c.r.s. round.*

The connecting links . . .

Two connecting links are needed. They are a matched pair made from 3/8" x 3/4" aluminum flat bar as shown in figure 140. Each link is bushed on one end with a .626" O.D. x 1/2" I.D. x 1/2" long bronze bushing. The link pin shown in figure 141 is pressed into the other end of each link. See figure 142.

When the bushings have been pressed in, drill a 1/16" hole through to the 1/2" bore of each one. Approximate location of the oil holes is shown in figure 142A. Countersink the oil holes slightly to provide a pocket for the oil.

DRILL & REAM $\frac{5}{8}$ DIAMETER

$2\frac{5}{16}$

DRILL & REAM $\frac{3}{8}$ DIAMETER

$\frac{3}{8}$

$\frac{3}{4}$

R$\frac{3}{8}$

R$\frac{3}{8}$

$1\frac{9}{16}$

Figure 140 *Connecting link. Make two from 1/2" x 3/4" flat*

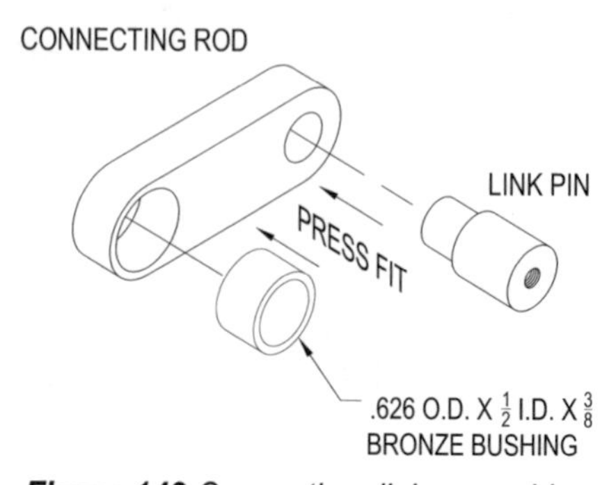

Figure 142 *Connecting link assembly.*

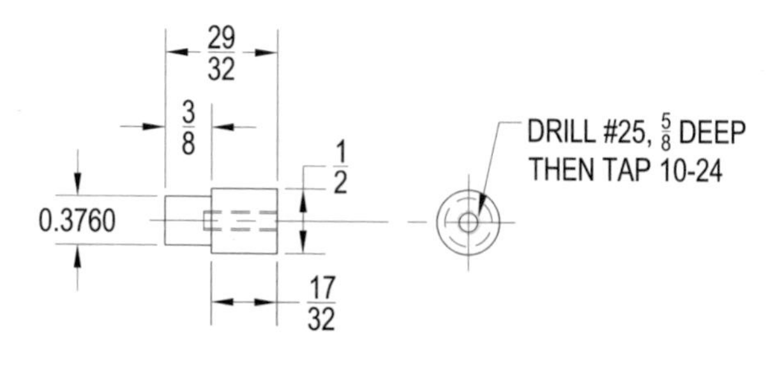

Figure 141 *Link pin. Make 2 from 1/2" diameter c.r.s. round*

Figure 142A *Drill a 1/16" oil hole in each connecting rod through to the 1/2" bore. Countersink the oil hole to provide a cup for the oil.*

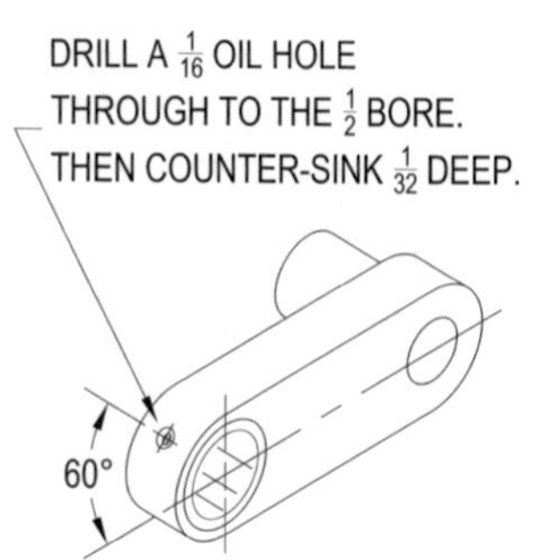

Oscillating Arms . . .

Begin the layout of each oscillating arm by first locating and punching the center of the crank boss. Next find and mark a vertical center line on the pivot boss. Locate and mark the hole location on pivot boss along the vertical center line 2-7/8" from the center of the crank boss. Find and mark a horizontal center line on the piston rod pivot boss. Then locate and mark the hole location on the piston rod pivot boss on the line 2-7/16" from the center of the pivot boss.

Aluminum makes for a satisfactory bushing material and since the oscillating arms are made of aluminum you could simply ream the crank and pivot bosses in each arm to shaft size and run them without bushings. In fact we did so on our first engines. But for better performance and greater longevity I think it best to go ahead and use bronze bushings.

To set up for drilling and reaming, clamp each arm in position on the mill/drill table. You can use parallel bars as spacers to raise the arm off the table. Straight holes are important so ensure the bosses are perpendicular to the drill spindle. I found it best to drill and ream each hole to size before moving to the next. Starting with the crank boss, step drill to 39/64", then ream to 5/8". Next, step drill the pivot boss to 39/64", then ream to 5/8". Finally, step drill the piston rod pivot hole to 19/64", then ream to .3135".

Locate, drill #29 then tap the hole in pivot boss located on back side of each oscillating arm for 8-32 threads. See figure 143.

Prepare a 5/8" arbor on the lathe as shown in figure 145.

Mount an oscillating arm on the arbor to begin work on the rear side. Secure with the

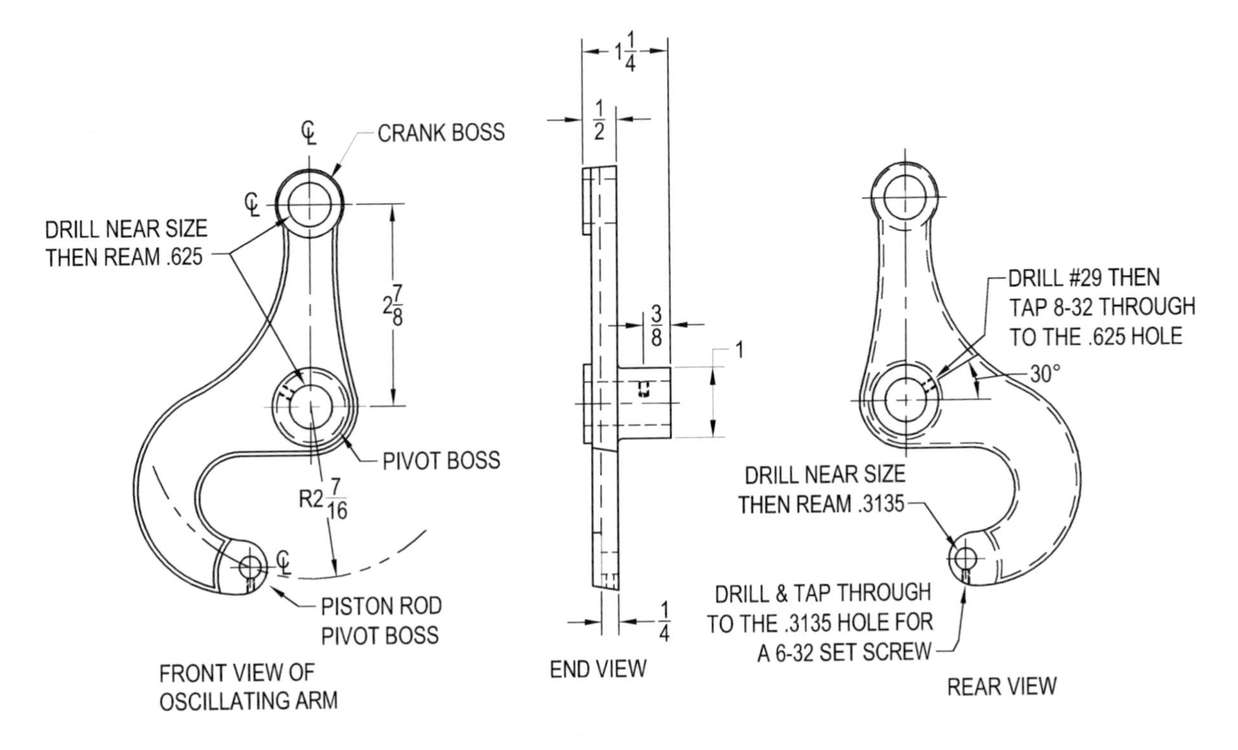

Figure 143 *Oscillating arms. Need an opposing pair.*

set screw. Face off the pivot boss to just slightly over 1-1/4" thickness and reduce the diameter of the boss to 1". Face off the remaining flat surface on the backside of the arm to remove the rough surface.

Reverse the arm on the arbor to face off the front side bosses. The crank boss to 1/2" thickness and the pivot should just clean up at 1-1/4" thickness.

Remove the arm from the arbor and mill the front side of the piston rod pivot boss to 1/4" thickness. Then cross drill and tap the piston rod pivot for the 6-32 set screw.

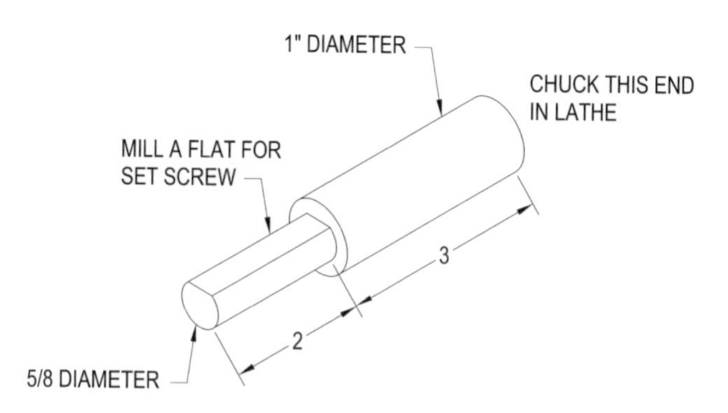

Figure 144 *Work arbor for machining the oscillating arms.*

Press 1/2" I.D. x 5/8" O.D. x 1/2" long bronze bushings into each crank bore and 1/2" I.D. x .626" O.D. x 1-1/4" bronze bushings in each pivot bore. See figure 145.

It will be necessary to lubricate the bushings from time to time so drill 1/16" oil holes through to the crank and pivot bores where shown in the figure. Countersink each oil hole to form a slight cup for the oil.

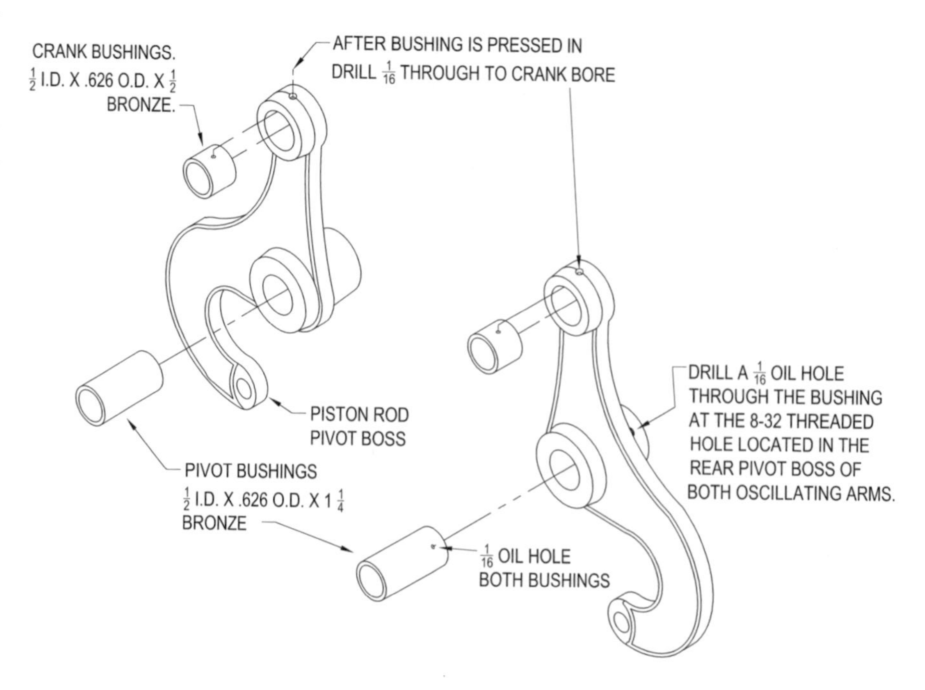

Figure 145 *After machining the oscillating arms, press bronze bushings into the pivot & crank bores. Then drill a 1/16" oil hole through to the crank and pivot bores of each arm where shown in the drawing. Note* Center the 1/16" oil hole for the crank bore in the existing 8-32 threaded hole in the crank boss.*

Milling the cylinder ends . . .

Before mounting the cylinder to the front panel, mill out the ends as shown in figure 146. This is necessary to give clearance for the oscillating arms and piston rods. Be very careful not to damage the cylinder. When finished, clean the sharp edges left by the milling cutter with a 3 corned scraper.

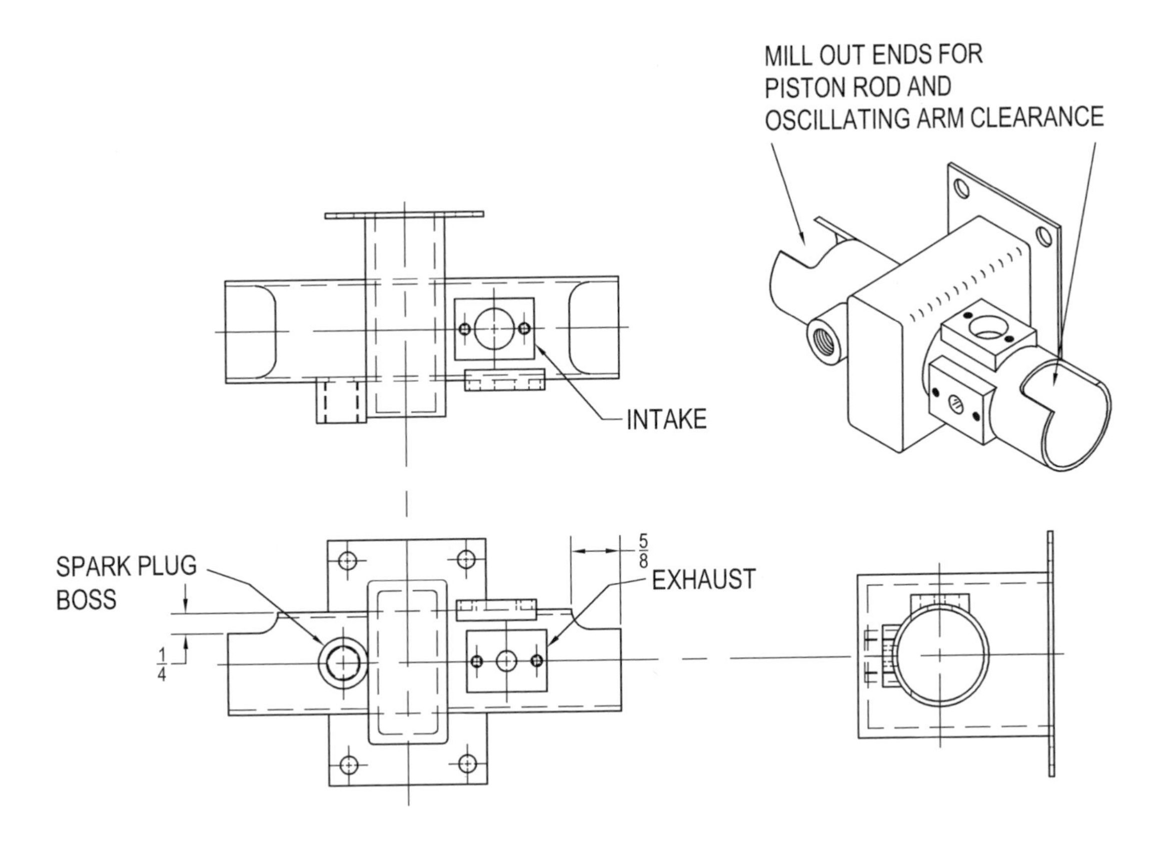

Figure 146 *Mill out both ends of cylinder for piston rod and oscillating arm clearance.*

Mounting & positioning the cylinder . . .

Note the drawing showing the assembly of the oscillating arms and connecting rods in figure 151. It shows the cylinder temporarily held in position with a bar clamp. The photo in figure 147 shows the type of clamp I used to hold the cylinder in position.

You will need a 1/2" I.D. spacer bushing for the linkage assembly as shown in figure 148. Make it now from 5/8" diameter aluminum or brass.

Preliminary measurements for locating the cylinder on the front panel are as follows. This will likely not be the final position of the cylinder, but it is a starting point.

First find the vertical center of the front panel and scribe a line. Measure over 1/8" to the right of the center line and scribe another

Figure 147 *This is the furniture clamp I used for positioning the cylinder.*

line parallel to the center line. Find and mark a vertical center line on the cylinder water jacket mounting plate.

Facing the front of the engine, position the cylinder so the center line on the base of the water jacket is aligned with the line scribed 1/8" to the right of center on the front panel. Place the main body of the engine on a flat surface. Position the cylinder on the front panel mounting pad so that a measurement taken from the center of each end of the cylinder bore to surface of the bench is 2-3/4". Refer to figure 149. With the cylinder in position, clamp it to the front panel as shown in figure 150.

With the cylinder clamped in position, insert a piston with piston rod attached into each end of the cylinder as shown in figure 150. Place one of the connecting links on the link pin as shown in the assembly drawing figure 151. Place the left side oscillating arm on the left side pivot shaft. Fit the crank bore end of the oscillating arm on the link pin. Place the right side oscillating arm on the right side pivot shaft. Install the other connecting link as shown in the figure. Screw in 10-24 x 3/8" socket head cap screws with #10 washers where shown to secure the linkage. Position the piston pivot hole at the end of each oscillating arm in the piston rod yoke. Align the .3135" holes in the piston rod and yoke, insert the piston rod pins which are made as shown in the upper left hand corner of figure 152. Then secure the pins with 6-32 x 3/8" set screws.

With the cylinder clamped in its temporary position and all the other above components installed you can locate the proper position the cylinder. Working from the front of the engine, rotate the flywheel in a clockwise direction. What we are after is the following:

With the engine just beyond ignition and about to begin the power stroke, the end of the left side piston should be aligned as close as possible with the left end of the cylinder.

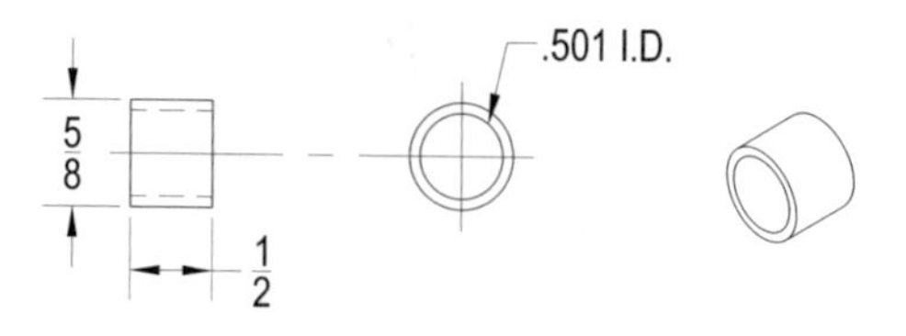

***Figure 148** Spacer bushing. Make one from 5/8" diameter aluminum of brass.*

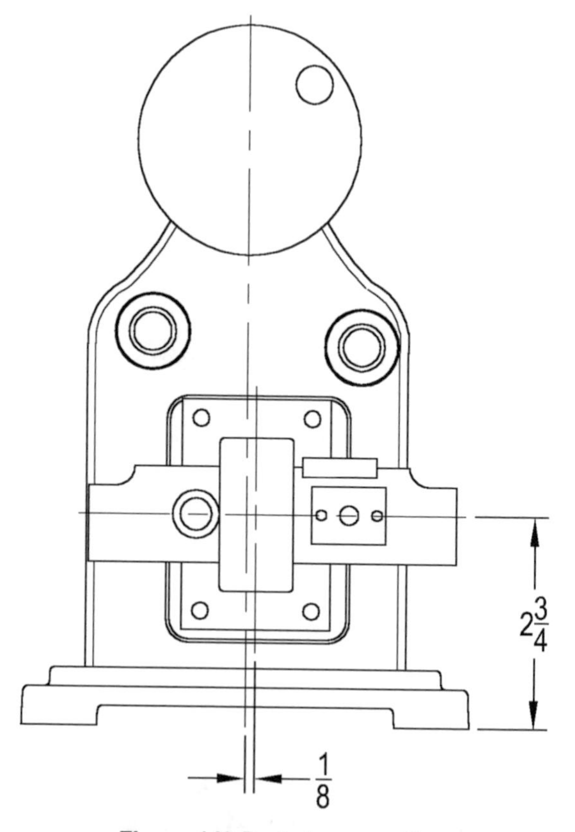

***Figure 149** Preliminary position of cylinder on front panel.*

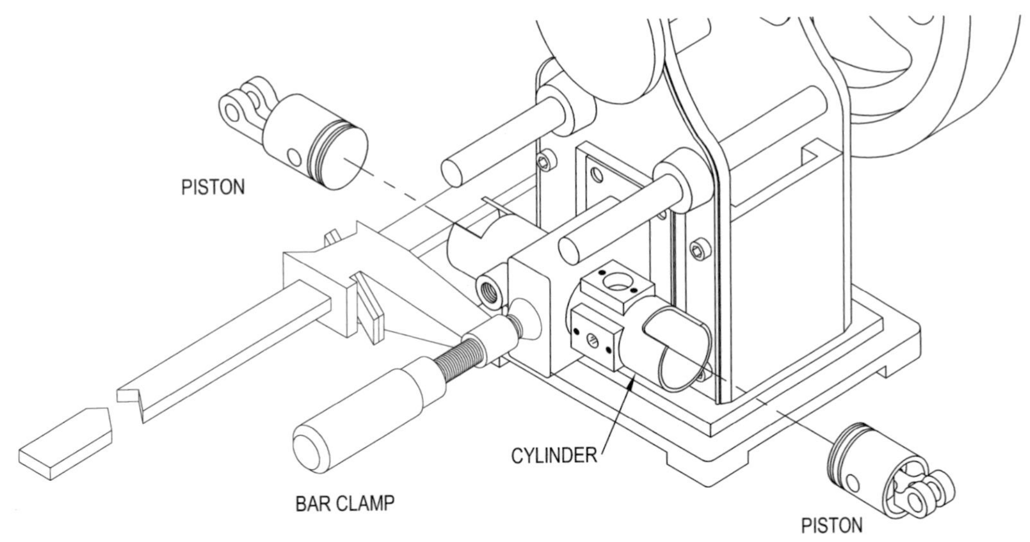

Figure 150 *Insert a piston into each end of the cylinder.*

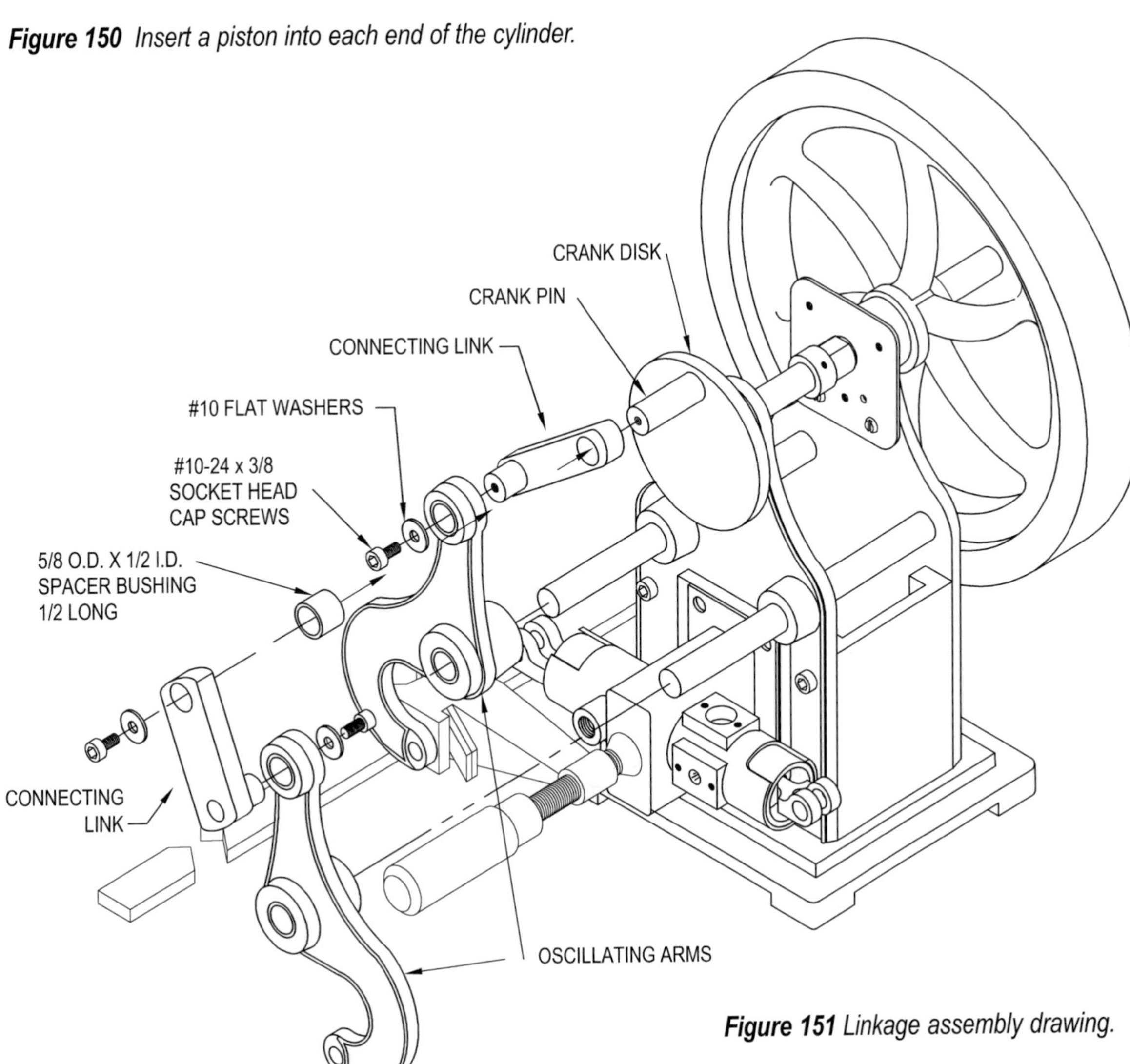

Figure 151 *Linkage assembly drawing.*

See the photo in figure 153A. Then when the engine reaches the end of the exhaust stroke, the end of the right piston should be positioned to align as close as possible with the right end of the cylinder. See photo in figure 155A.

You will likely have to reposition the cylinder several times before the optimum position is finally reached. This means raising, lowering or moving the cylinder from side to side in order to find the right combination.

The next order of business is to check the final position of the pistons in reference to the exhaust, intake and spark plug holes.

The drawings in figures 153, 155, and 156 show the position of the pistons at power, exhaust and intake. The important thing is to provide an open port at ignition, an open port at exhaust and an open port at intake. By open port I mean the respective holes in the cylinder wall are not totally blocked by the pistons during ignition, intake and exhaust.

Note: On the engines we have built the final length of connecting links shown in figure 140 has consistently been 1-9/16" between holes center to center. You could experience problems arriving at an optimum position for the cylinder, or the pistons may not position properly in reference to the intake, exhaust and spark plug holes at each cycle as they should. The pistons may even hit each other during travel. If this happens, it is not a serious problem and could have been caused by

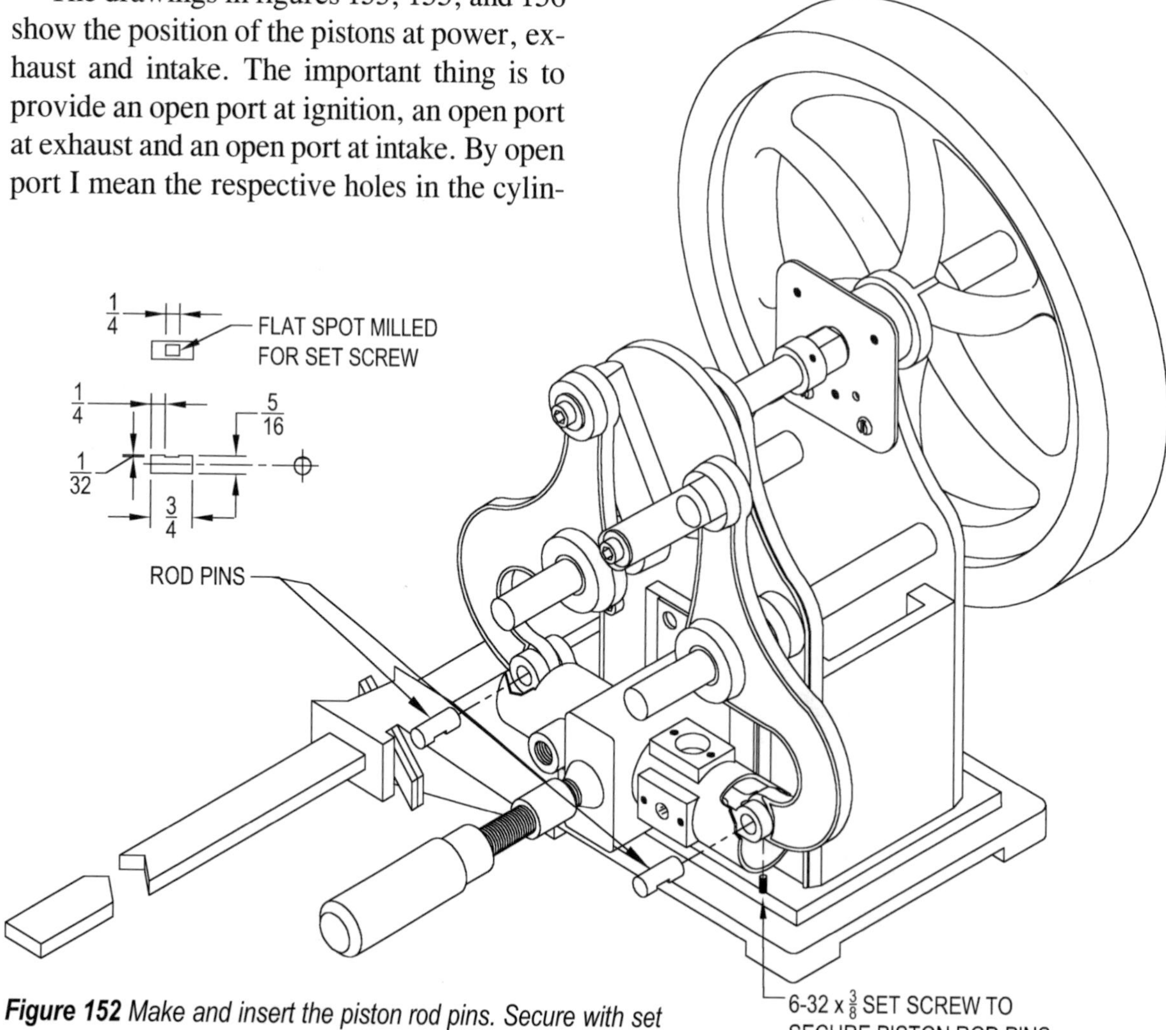

Figure 152 *Make and insert the piston rod pins. Secure with set screws. Piston rods are made from 5/16" diameter c.r.s. round.*

slight differences in layout as you built your engine. The solution would be to make a new shorter set of links. Say 1-1/2" between centers.

When satisfied with the position of the cylinder, use a transfer punch to transfer the hole locations from the cylinder mounting plate to the front panel. Disassemble the linkage and remove the cylinder and then the front panel

END OF IGNITION
BEGINNING OF POWER
1/8
LEFT PISTON
RIGHT PISTON
SPARK PLUG HOLE IN CYLINDER WALL

Figure 153 *This drawing shows the position of the pistons in reference to the spark plug hole when the engine is at the beginning of the power stroke. Looking through the spark plug hole, this is how the pistons should be positioned in the cylinder. Note: Both pistons have cleared the spark plug hole in the cylinder wall in order to allow the spark to ignite the fuel mixture. See photo figure 153A to see how the oscillating arms and linkage are positioned at this part of the cycle..*

from the main frame to drill and tap the four mounting holes in the front panel for 8-32 threads.

Now would be a good time to dissasemble the engine and give it a paint job. I painted the main body of my engine green and the oscillating arms and flywheel were painted red. The cylinder, and links were painted black. After the paint has dried, reassemble the main body and mount the cylinder. Upon reassembly of the main body of the engine remember to apply a bead of silicon gasket sealer along the joints between the panels. The reason being that water or antifreeze will be held inside the main body of the engine for cooling purposes and sealing the joints will prevent leaks.

We'll be working on the valves next and when mounting the intake valve, the right side pivot shaft will get in the way. So as you reassemble the engine, just leave it off for the

Figure 153 A *The engine at the beginning of the power stroke. The ignition points opened just prior to this and the fuel mixture has been ignited. Note the position of the piston on the left. Its back edge is aligned very near the left end of the cylinder.*

Figure 154 *This is a photo of the engine showing the position of the linkage at the end of expansion.*

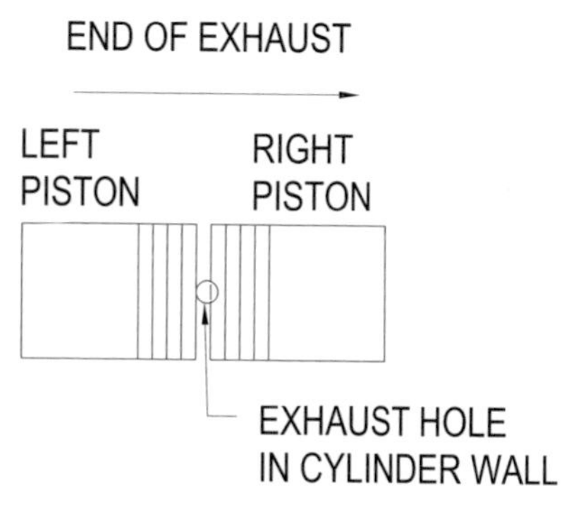

Figure 155 *Here the engine is at the end of exhaust and at the beginning of intake. Looking through the exhaust hole, this is how the pistons should be positioned in the cylinder. Note the left piston is at the edge of the exhaust hole in the cylinder. The right piston is far enough over to open the exhaust port, but still covers some of the hole. Refer to the photo in figure 155A for the corresponding position of the oscillating arms and linkage.*

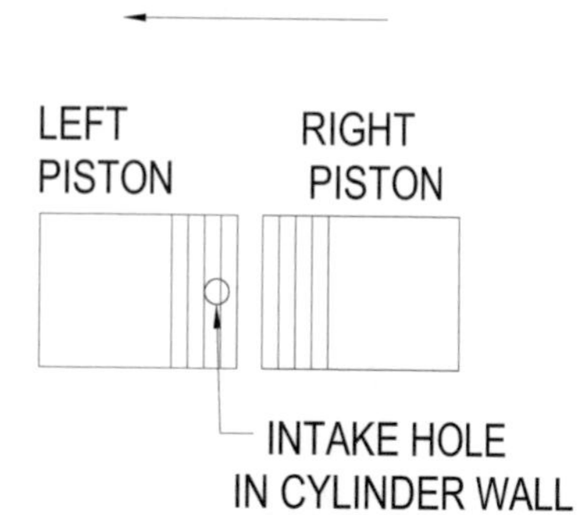

Figure 156 *This is a look through the intake hole when the engine is at the end of the exhaust stroke and at the beginning of intake. Looking through the intake hole, this is how the pistons should be positioned in the cylinder. Note the how the left piston is covering the intake hole in the cylinder. This quickly changes though. As the flywheel rotates, the left piston rapidly picks up speed and moves away from the right piston opening the intake port and sucking in the fuel. The right piston follows and covers the intake port which begins the compression stroke. Refer to the photo in figure 156A for the corresponding position of the oscillating arms and linkage.*

Figure 155 A *The engine is at the end of exhaust and the beginning of intake. Notice that the end of the right piston is aligned very near the right end of the cylinder.*

Figure 156 A *This is a photo of the engine showing the position of the linkage at intake.*

time being. The cylinder is mounted to the front panel with four 8-32 x 3/4" socket head cap screws as shown in figure 157. As shown in the figure, make a gasket to fit between the cylinder mounting plate and front panel and for added leak protection coat the gasket with sealer. The linkage will be assembled later after the valves have been constructed and mounted to the cylinder.

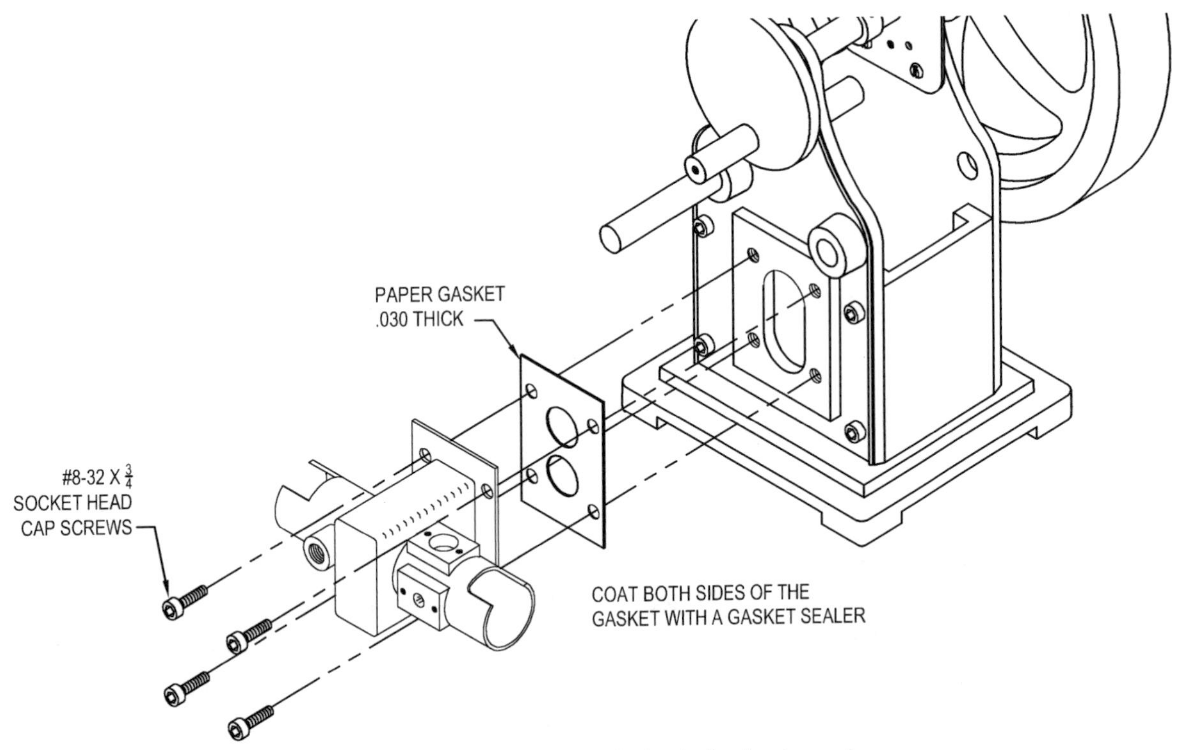

Figure 157 *Bolt the cylinder to the front panel*

Making the exhaust valve . . .

Both valves are simple and self activating with no cams or linkage necessary. Each valve assembly is contained in a separate block that is mounted to the cylinder.

Start by squaring a 3/4" x 3/4" x 1" block of aluminum for the exhaust valve body shown in figure 158. Prepare the block by locating the center of one side. Step drill through the center to 27/64" then ream to .4385". Then locate, drill & tap the two 6-32 holes 3/8" deep and locate and drill the two 9/64" holes through.

Make the exhaust valve guide as shown in figure 159 by chucking a length of 1/2" diameter c.r.s. round rod in a collet so that at least 1-1/4" sticks out past the chuck. Reduce the outside diameter to .4395", 1" back from the end. Center drill, then drill

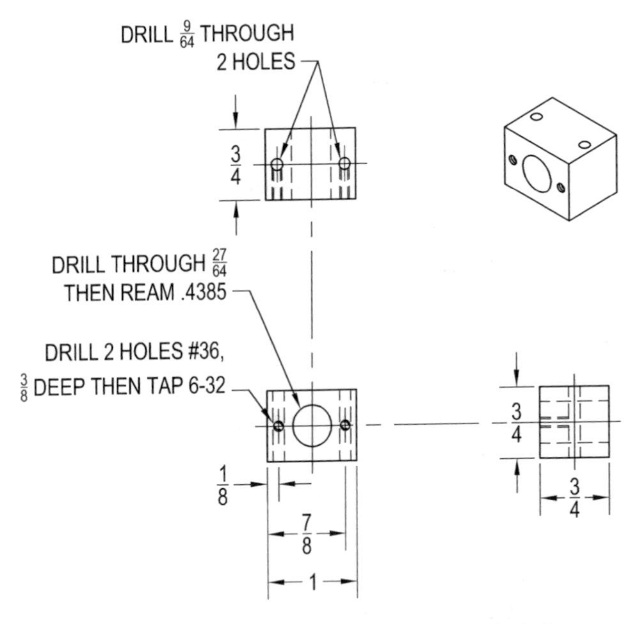

Figure 158 *Exhaust valve body. Made from 3/4" x 3/4" aluminum bar stock..*

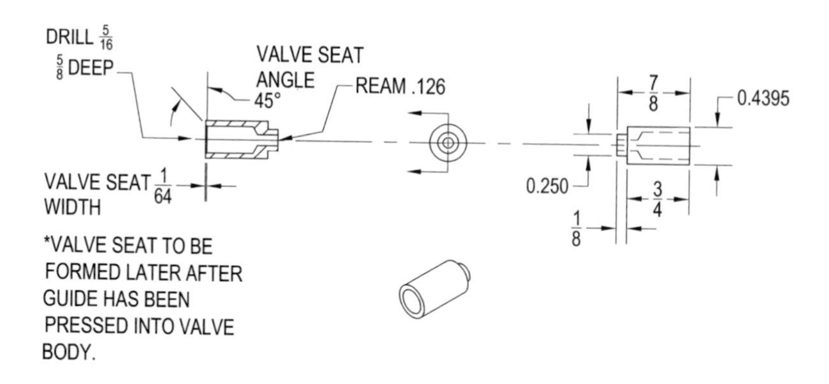

Figure 159 *Exhaust valve guide. Make from 1/2" diameter c.r.s. round rod.*

and ream to .126", 1" deep. Then increase the size of the .126" hole to 5/16", 5/8" deep. Use a parting tool to turn the 1/4" diameter shoulder then part off to a total length of 7/8". The valve seat shown in the drawing will be

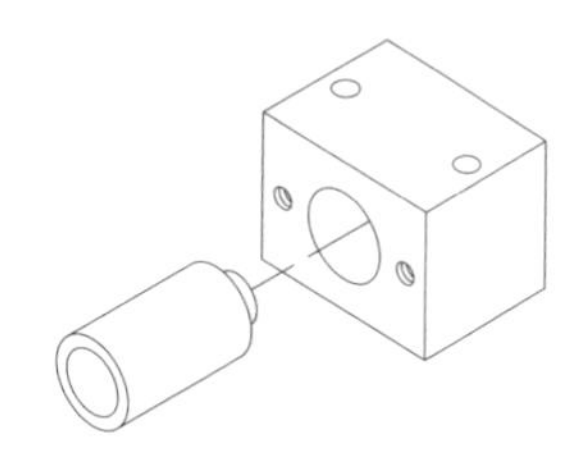

Figure 160 *Press the valve guide into the valve body.*

formed later after the guide has been pressed in the valve body. See "Valve seat forming" on page 92.

The valve guide is ready to press in the valve body, but first coat the outside of the guide with epoxy, then press it into the valve block as shown in figure 160. The reason for the epoxy is to seal potential air leaks around the guide which could affect the engine compression. The seat end of the valve guide should be recessed in the valve body approximately 1/64". The neck end of the valve guide will extend out from the body 1/8". Drill the 7/32" exhaust port hole through the side of the valve body and into the inside bore of the valve guide as shown in the figure 161.

Make the exhaust valve stem as shown in figure 162 from a piece of 1/8" drill rod 1-1/4" long. You can use a "V" block to hold the stem in position while drilling the 1/16" hole at the end.

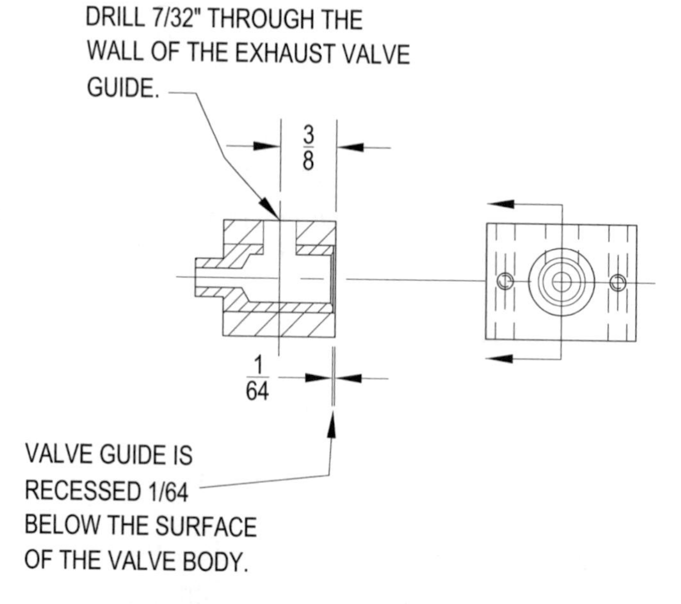

Figure 161 *Drilling the exhaust port.*

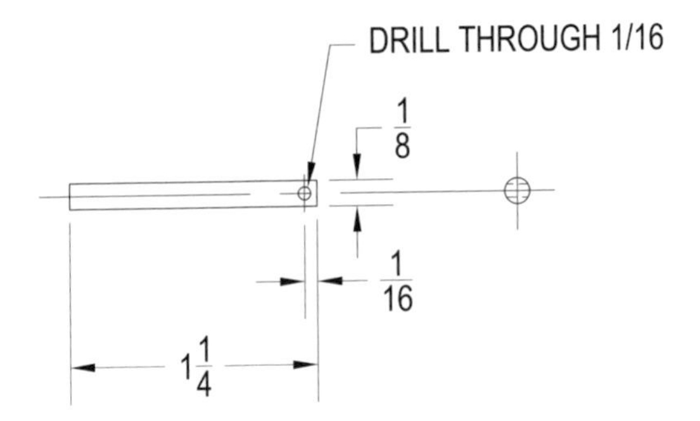

Figure 162 *Exhaust valve stem. Make from 1/8" diameter c.r.s. round rod.*

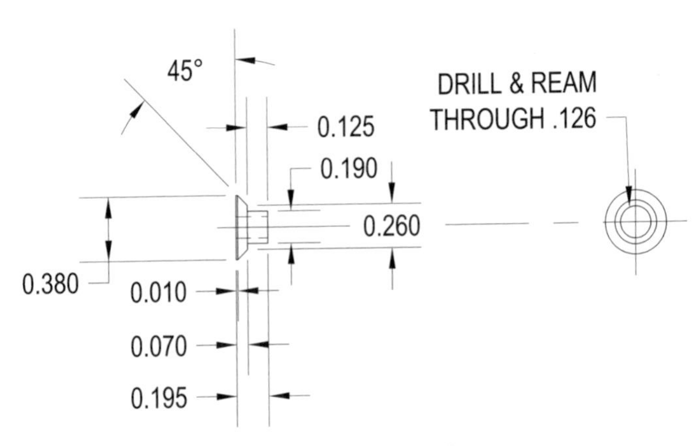

Figure 163 *Valve head. Make two from 7/16" diameter c.r.s. round rod.*

The valve heads for the exhaust and the intake valves as shown in figure 163 will be the same so you can make two of them now. To make the valve heads, chuck a short length of 7/16" round rod in a collet. Face off, drill and ream through .126". Turn a .190" diameter shoulder on the end 1/8" back. Adjust the compound rest to turn the 45 degree bevel. Then cut off the valve head long enough to allow for facing off. Repeat to make the second valve head.

Insert the valve stem into the .126" reamed hole in one of the valve heads positioning the end of the stem flush with the head of the valve. See figure 164 and 165. When in position braze the valve head to the stem.

Then, mount the valve stem in a 1/8" collet to face off the valve head and true the 45 degree taper. The finished thickness of the .380" diameter section of the head to be .010".

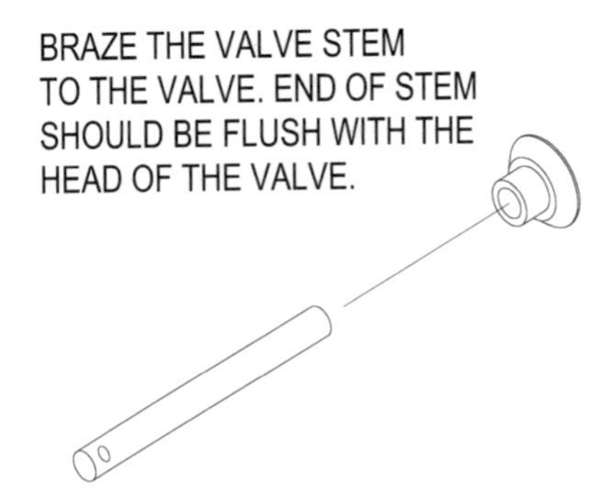

Figure 164 *Position the valve head on the end of the valve stem. The end of the stem should be flush with the top surface of the valve head.*

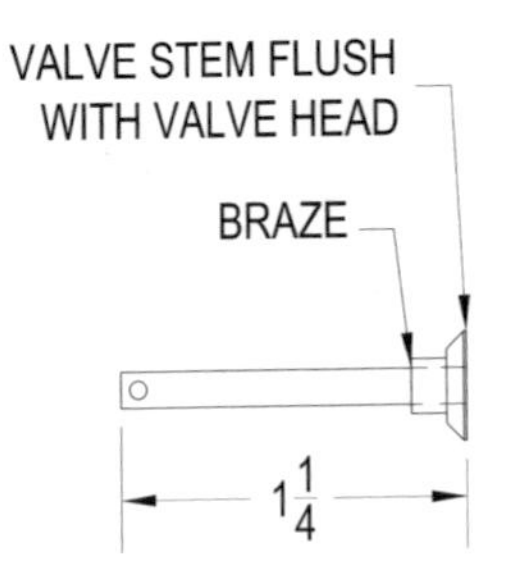

Figure 165 *Side view showing the valve stem flush with the valve head.*

Valve seat forming . . .

A special tool can be made to form the valve seat. It's a simple tool and one used in our last project when we built the Atkinson Cycle engine.

The valve seat tool and seat narrowing tool are made of aluminum and then charged with lapping compound. The seat tools are shown in figure 166 and 167 and the setup is shown in figure 168.

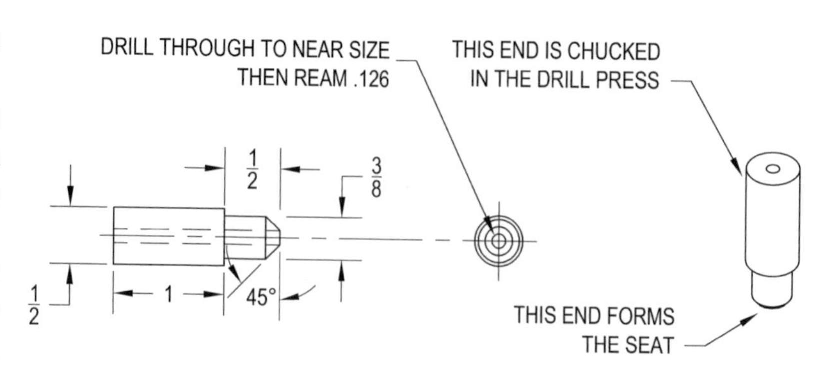

Figure 166 Valve seat forming tool. Made from aluminum round.

To form the seat, charge the end of the seat tool with valve grinding compound and chuck it in a drill press. Insert the pilot shaft through the valve guide and into the valve seat forming tool. The purpose of the pilot shaft is to keep the valve guide in perfect register with the tool. While running the drill press at a slow to moderate speed force the valve guide against the seat tool giving it a 1/4 turn every minute or so until the width of the seat measures about 1/64". During the process be sure and hold the end of the pilot shaft to keep it from spinning in the valve guide and enlarging the guide hole. If you over do it and grind the seat too wide you can narrow it with the narrowing tool which has a flat end.

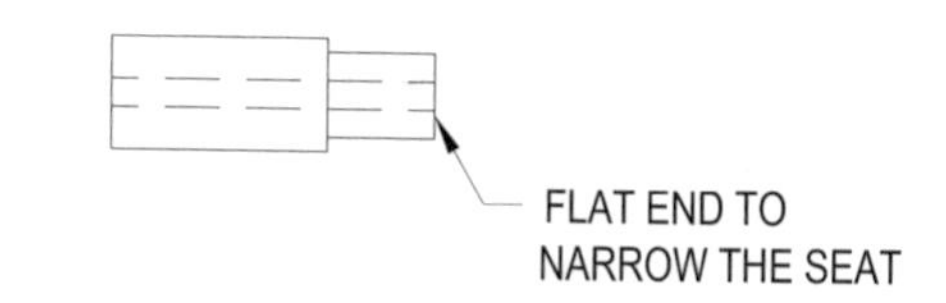

Figure 167 Valve seat narrowing tool. Also made from aluminum round. Notice that this tool is the same as the seat forming tool, but with a flat end.

When the seat has been formed, lap the valves with valve grinding compound. One way to do the lapping is to insert the valve into the valve guide and hold its stem

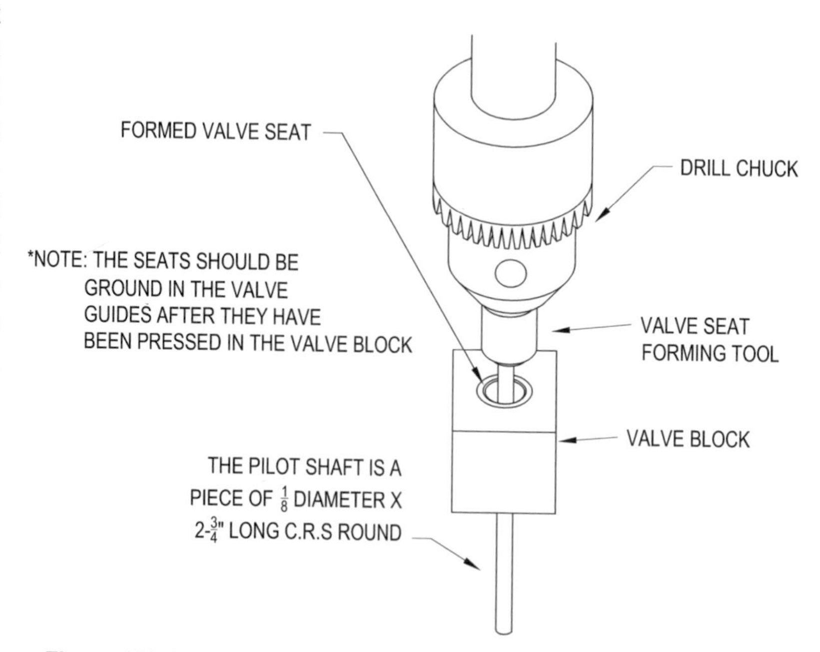

Figure 168 The set up for grinding the valves. It is important that the valve seats be ground after the valve guides have been pressed into the valve blocks.

with a pin vise. Apply fine grit lapping compound and twist the valve back and forth against its seat rotating a 1/4 turn every fourth or fifth twist. You can tell when the valve has been lapped when upon close examination you can see a narrow, uniform gray line around the surface of the seat and the valve. Be sure and mark the valves in such a way that you will know which seat they are made for.

The valve spring retainers are shown in figure 170. Two are required. To make them, chuck a piece of 3/8" diameter c.r.s. round rod in a 3 jaw chuck. Center drill, then drill 7/64", 1" deep. Ream the 7/64" hole with a .126" reamer. Counter bore to 1/4" diameter, 1/32" deep. Use the parting tool to turn the 1/4" diameter shoulder then part off 1/8" long. Repeat the process to make the second retainer.

The spring used for the valves was purchased at the local hardware store. A photo and detailed information on the spring can be seen in figure 169. Cut two valve springs from the 1-1/2" long spring shown in the figure. Each spring to be 4 coils long.

To assemble each valve, insert a retainer into the top of a spring. Place the spring with

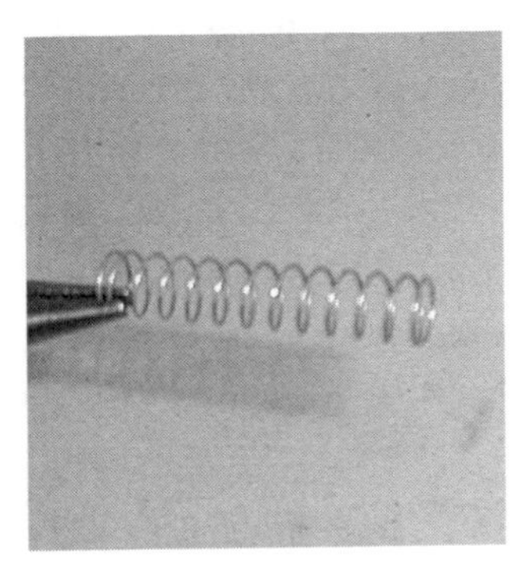

Figure 169 *Springs such as the one shown above can be purchased at most hardware stores and work well as valve springs. Both valve springs can be cut from this single 1-1/2" long spring which measures, .310" O.D. x .288" I.D. Wire thickness is .0210.*

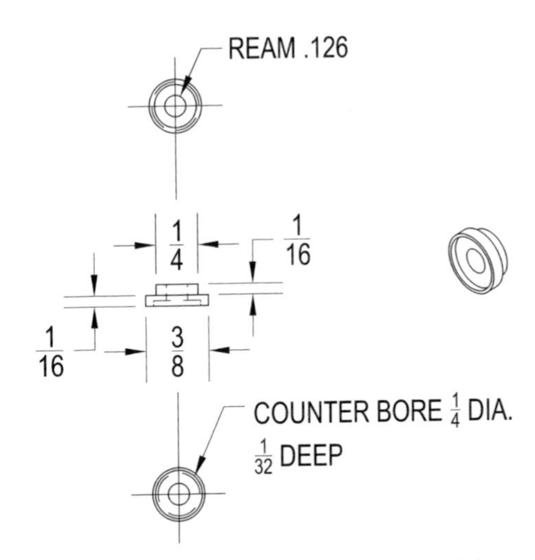

Figure 170 *Valve spring retainer. Make two from 3/8" diameter c.r.s. round rod.*

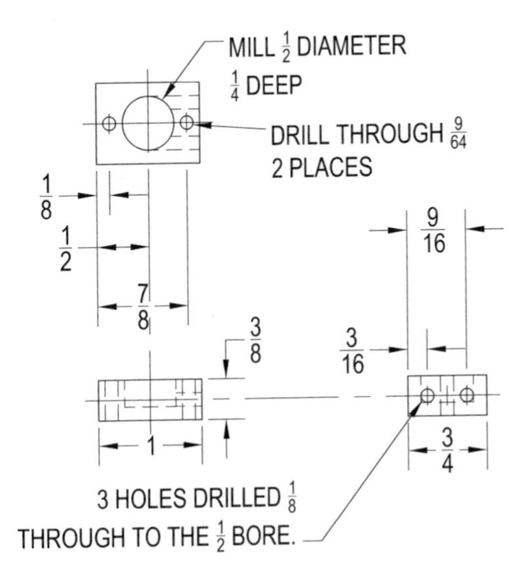

Figure 171 *Muffler/diverter. Make one from a 3/8" x 3/4" x 1" block of aluminum.*

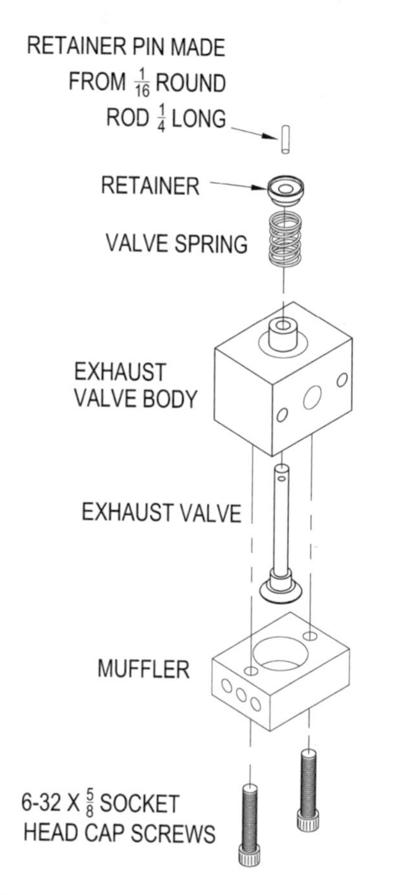

Figure 172 *Assembly drawing of the exhaust valve.*

retainer over the 1/4" diameter shoulder on one of the valve blocks. Insert the valve and compress the spring so the 1/16" hole in the valve stem extends out far enough to put the retainer pin in the hole.

When the spring is decompressed, the pin will rest inside the counter bore in the retainer and hold the valve assembly together. The exhaust valve assembly is shown in figure 172 and the intake valve assembly is shown in figure 181.

Make the muffler/diverter next as shown in figure 171. It will divert the exhaust out to one side through two 1/8" holes and at the same time keep the oil residue from spraying on to your clean shirts. It mounts on the valve block as shown in figure 172 with two 6-32 x 5/8" socket head cap screws.

Intake valve assembly . . .

The same basic procedures are used to construct the intake valve as was used to construct the exhaust valve. Construction details are shown in figures 173 through 181. The differences are that the intake valve body is smaller than the exhaust valve body and because of this the valve stems are a bit shorter too. And then there is the intake valve cap where the carburetor will mount. And note that a gasket is used between the valve cap and the valve body.

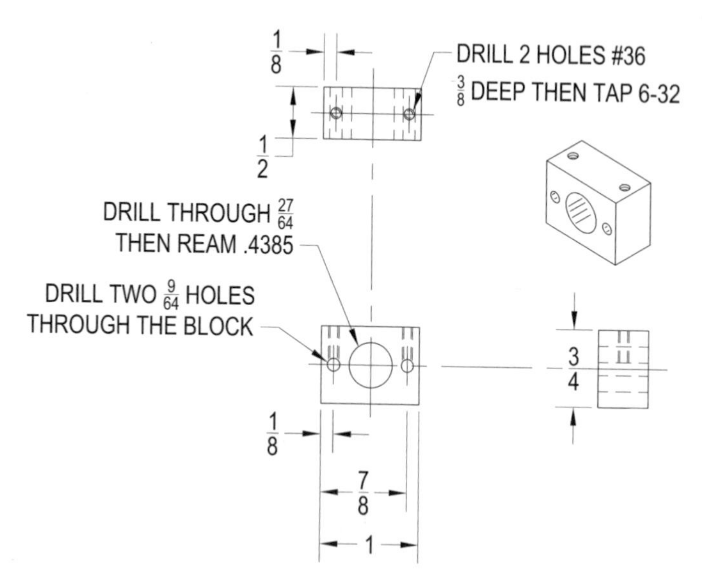

Figure 173 *Intake valve body. Made from a 1/2" x 3/4" x 1" aluminum block..*

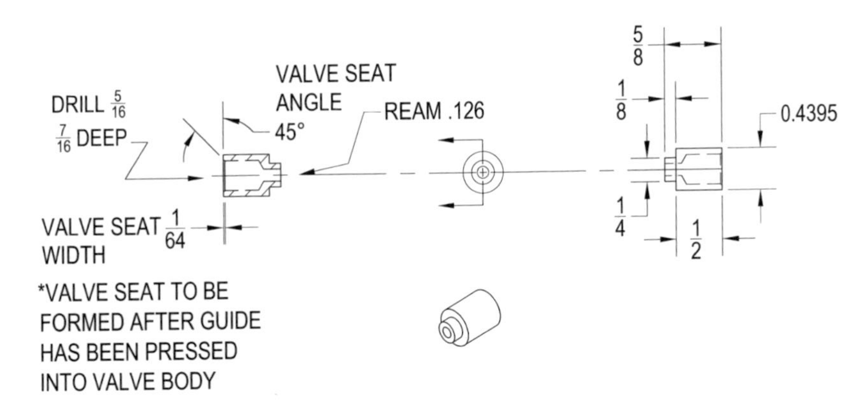

Figure 174 *Intake valve guide. Make from 1/2" diameter c.r.s. round rod.*

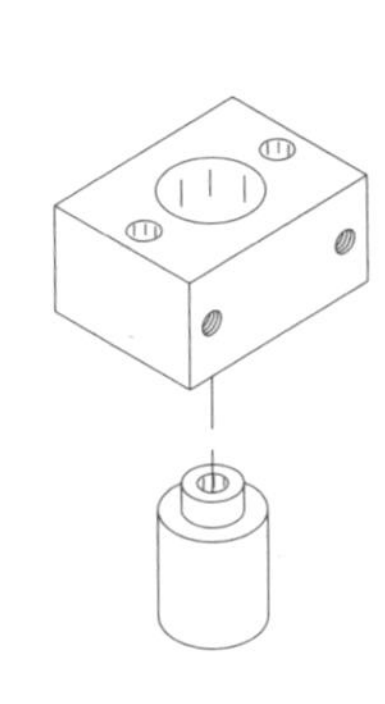

Figure 175 *Coat the valve guide with epoxy, then press it into the valve block.*

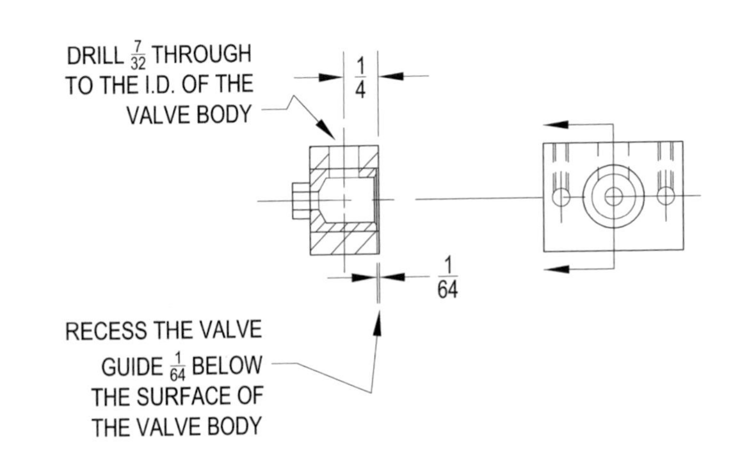

Figure 176 *Drilling the intake port*

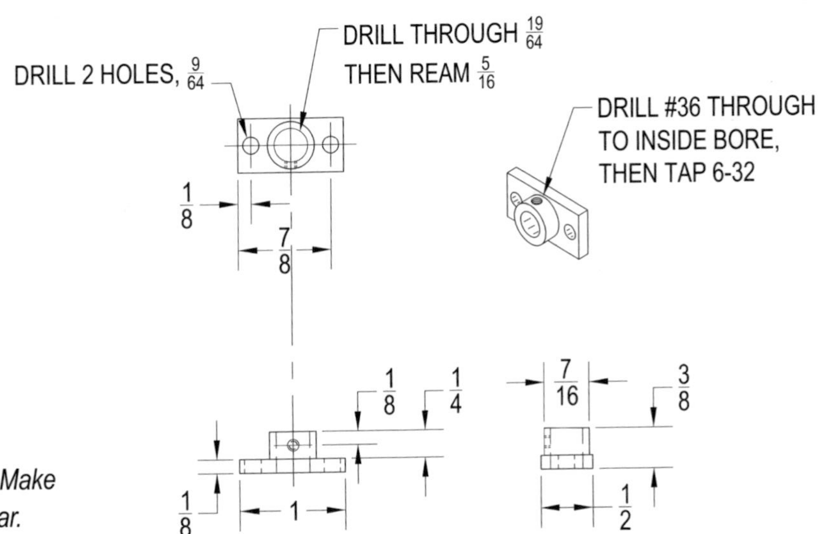

Figure 177 *Intake valve cap. Make 1 from a 1/2" x 1" aluminum bar.*

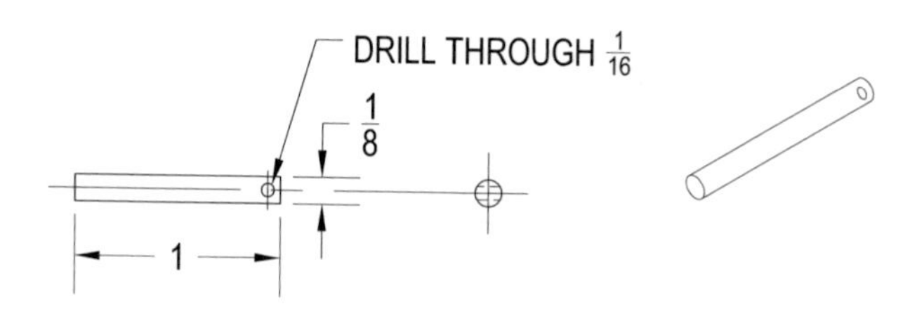

Figure 178 *Intake valve stem. Make one from 1/8" diameter c.r.s. round rod.*

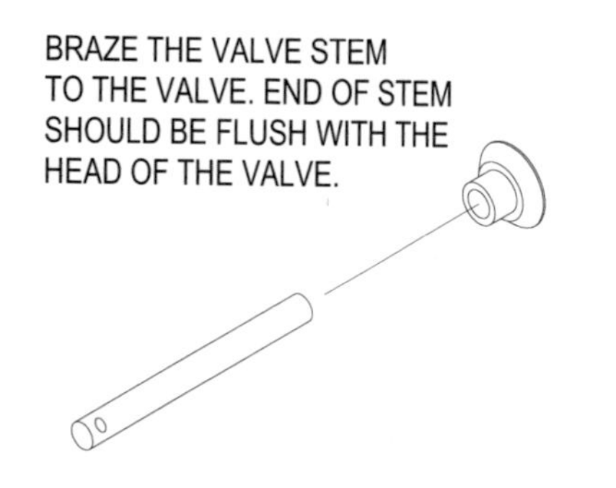

Figure 179 *Position the valve head on the end of the valve stem as shown in the drawing. The end of the stem should be flush with the head of the valve.*

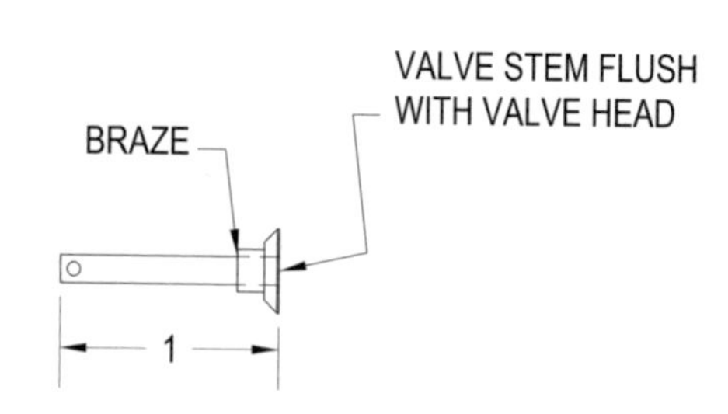

Figure 180 *Braze the valve stem to the valve head.*

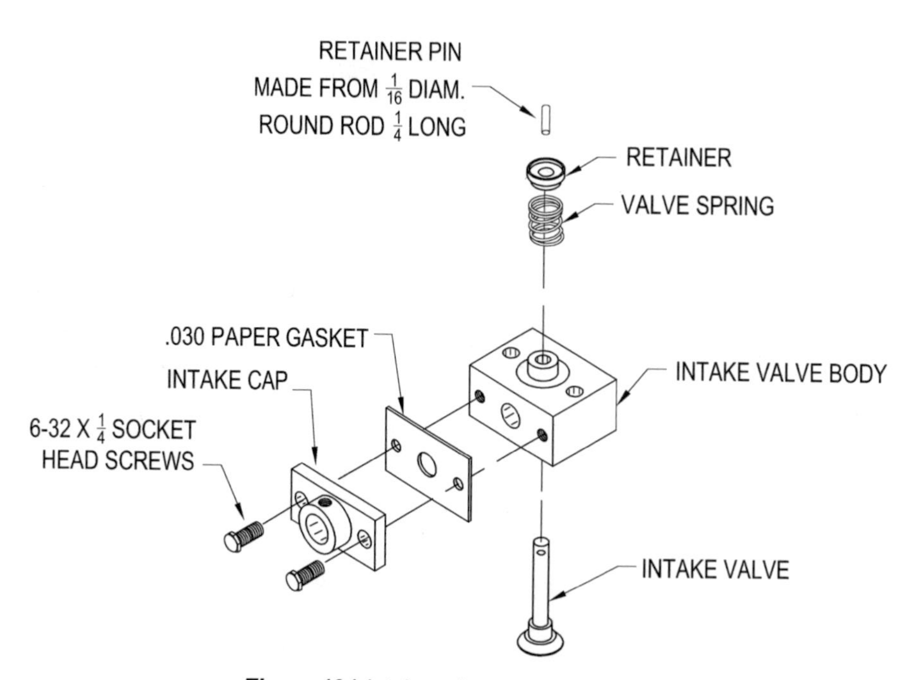

Figure 181 *Intake valve assembly*

Mounting the valves on the cylinder. . .

Figure 182 *Mounting the valves on the cylinder in their respective positions. As the drawing shows, gaskets are used between the valve bodies and cylinder mounting blocks. 6-32 x 7/8" socket head cap screws are used to attach the exhaust valve body and 6-32 x 5/8" socket head cap screws are used to fasten the intake valve body.*

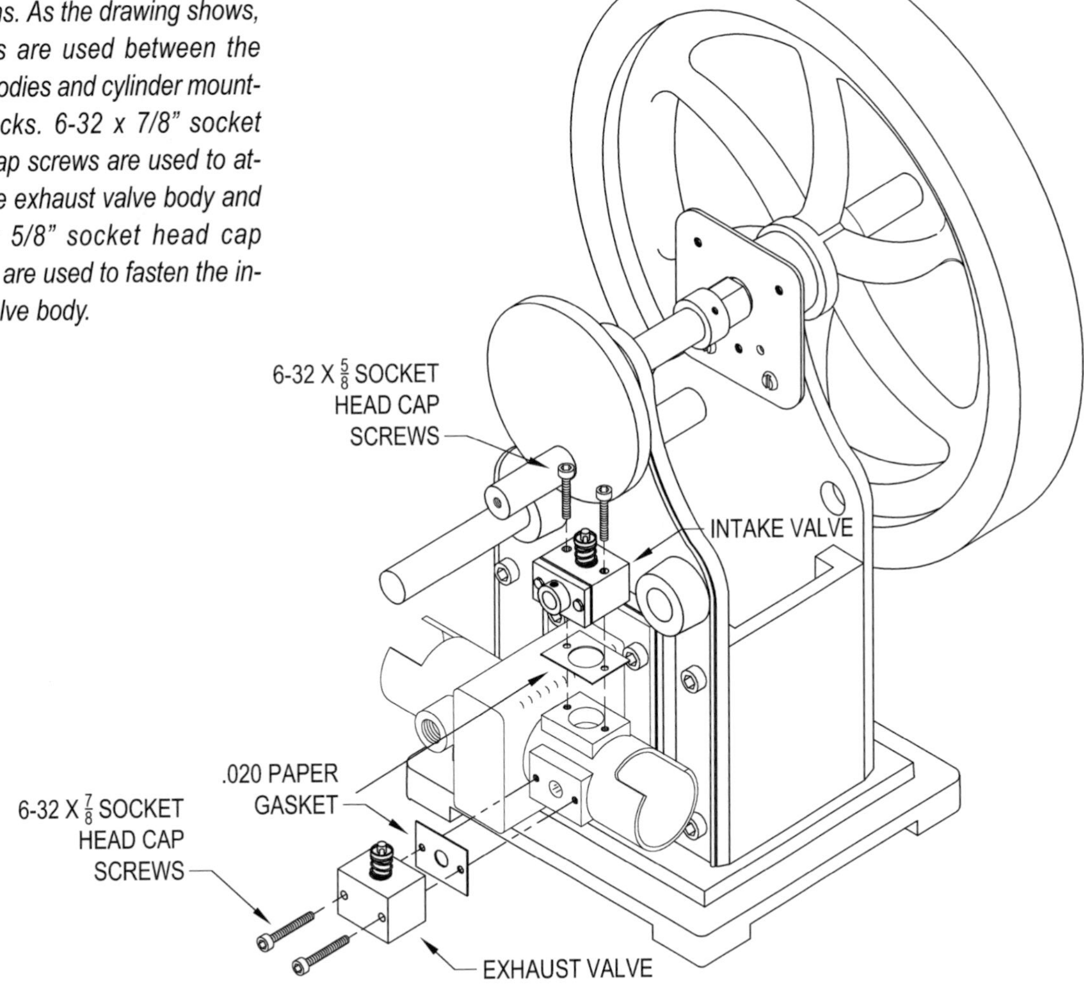

The carburetor . . .

The carburetor used is a very simple plain tube type. Air is pulled in through the front of the carburetor while fuel is pulled in through a needle valve jet located along its length. The fuel/air mixture takes place inside the tube and is pulled into the cylinder.

The parts of the carburetor are the main body, the air intake orifice and the needle valve jet assembly.

A drawing of the carburetor tube is in figure 183. To prepare it, chuck a length of 1/2" diameter aluminum or brass round in the lathe using the 3-jaw chuck. Center drill then drill through with a 1/4" drill. Turn the 5/16" diameter end 7/32" back. Use a 3/32" parting tool to turn the cooling fins along the length of the carburetor tube. Part off the carburetor at 1-1/8" long. Then clamp it in a "V" block to drill and tap the 6-32 holes on both sides of the tube. One of the 6-32 holes will be for the jet and the other for the needle valve.

The air intake orifice is shown in figure 184. It is also prepared from a length of 1/2" diameter aluminum or brass round chucked in the lathe using the 3-jaw chuck. Begin by center drilling, then drill with a 5/64" drill at least 1/2" deep. Flare the inlet with a 45 de-

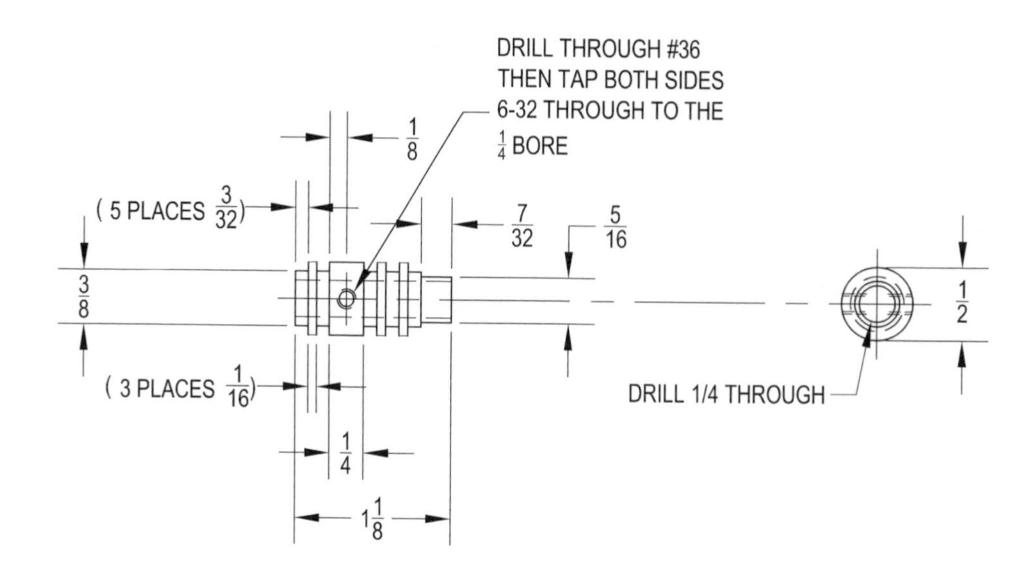

Figure 183 *Carburetor main body. Make one from 1/2" diameter aluminum round.*

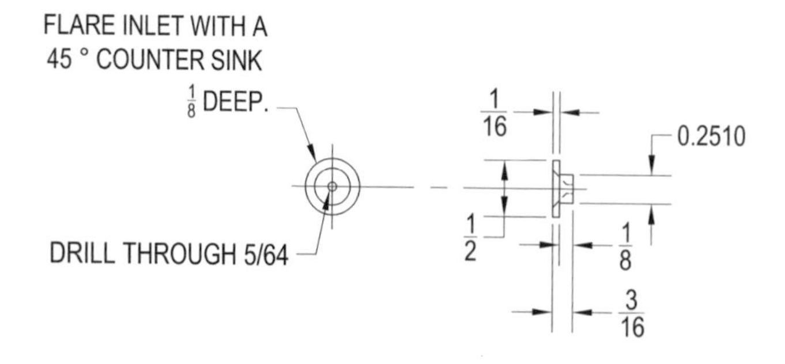

Figure 184 *Air intake orifice. Make one from 1/2" diameter aluminum round.*

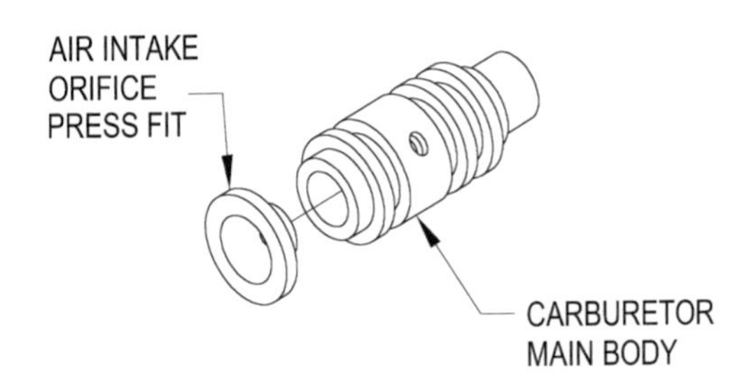

Figure 185 *Press fit air intake orifice into carburetor main body.*

gree, 1/2" counter sink 1/8" deep. Use the parting tool to turn the 1/8" wide .2510" diameter shoulder. Then part off at a total length of 3/16" as shown in the figure.

Prepare the jet as shown in figure 186 from a length of 3/16" brass hex rod held in the lathe with a 3-jaw chuck. Turn to .1365" diameter, 3/16" back from the end. Spot center and drill #73, 1/8" deep. Machine the 30 degree angle end leaving a slight flat for the #73 hole. Finally, thread 6-32 to the shoulder. In order to chuck the jet from the other end you will need to make an adapter that is threaded 6-32. The adapter can be a length of 1/2" diameter round held in the 3-jaw with a center hole tapped 6-32. Simply screw the jet into the adapter to drill the 1/16" hole through to the #73 hole. Then turn the end to .130" 1/4" back from the end. Finally, screw the jet into on of the 6-32 holes in the side of the carburetor tube.

The needle valve is made from an 6-32 x 3/4" brass screw, a #5 sharp sewing needle, and a knurled brass nut. See figure 187.

To make the knurled nut, chuck a piece of 3/8" diameter brass rod in the lathe. Apply a single straight knurl, medium or coarse. Drill and tap 6-32. Part off the knurled disc 3/32" thick.

Thread the knurled nut on the 6-32 x 3/4" brass screw. Soft solder the knurled disc to the screw head. Mount the screw in the threaded adapter used for making the jet. Face off the head of the screw to 1/32" Spot center and drill through the screw with a 1/16" drill. Screw the needle valve into the 6-32 hole in the side of the carburetor tube until it bottoms then back it out 3-1/2 turns. Slip a #5 sharp sewing needle into the 1/16" hole in the needle valve until it seats in the jet. Mark the needle with tape where it protrudes from the knurled head. Score the needle on the grinder and break off the excess length. We will be soldering the needle to the head of the knurled nut on the needle valve screw, but in order for the solder to stick to the needle we need rough up the blunt end with a piece of sand paper. Then reinstall the needle as shown in step 5 in figure 187. Apply flux and soft solder around the needle and into the head of the screw.

Remove the needle valve to install the tension spring and reassemble as in figure 188.

6-32 THREAD
DRILL $\frac{1}{16}$ $\frac{3}{8}$ DEEP
$\frac{1}{32}$
DRILL #73 THROUGH TO $\frac{3}{16}$ HOLE
0.130
$\frac{1}{4}$
30°
$\frac{3}{16}$ HEX
$\frac{1}{16}$
$\frac{1}{2}$

Figure 186 *The jet*

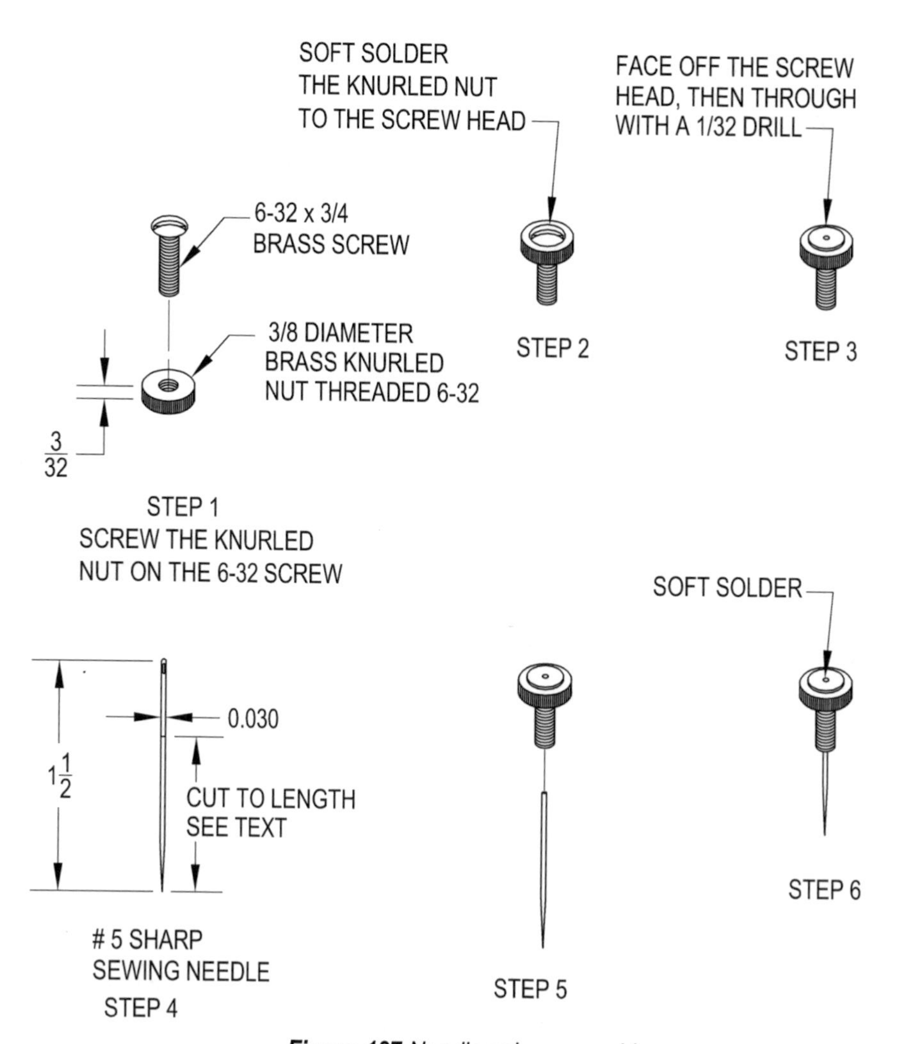

Figure 187 *Needle valve assembly.*

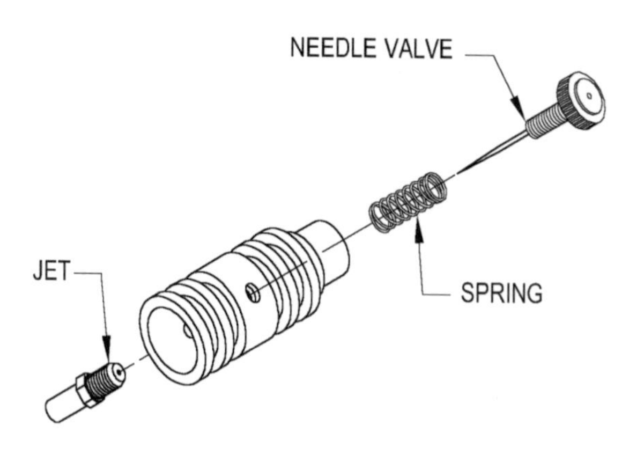

Figure 188 *Carburetor assembly. The needle valve spring shown came from a Retractable ink pen.*

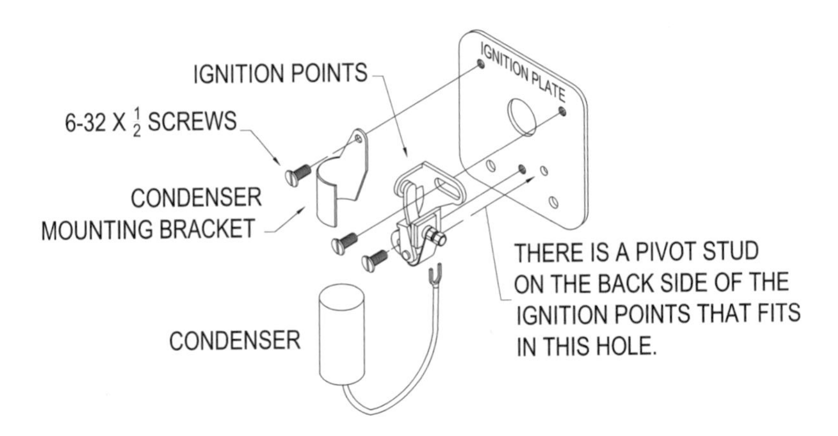
IGNITION POINTS
IGNITION PLATE
6-32 X $\frac{1}{2}$ SCREWS
CONDENSER
MOUNTING BRACKET
THERE IS A PIVOT STUD
ON THE BACK SIDE OF THE
IGNITION POINTS THAT FITS
IN THIS HOLE.
CONDENSER

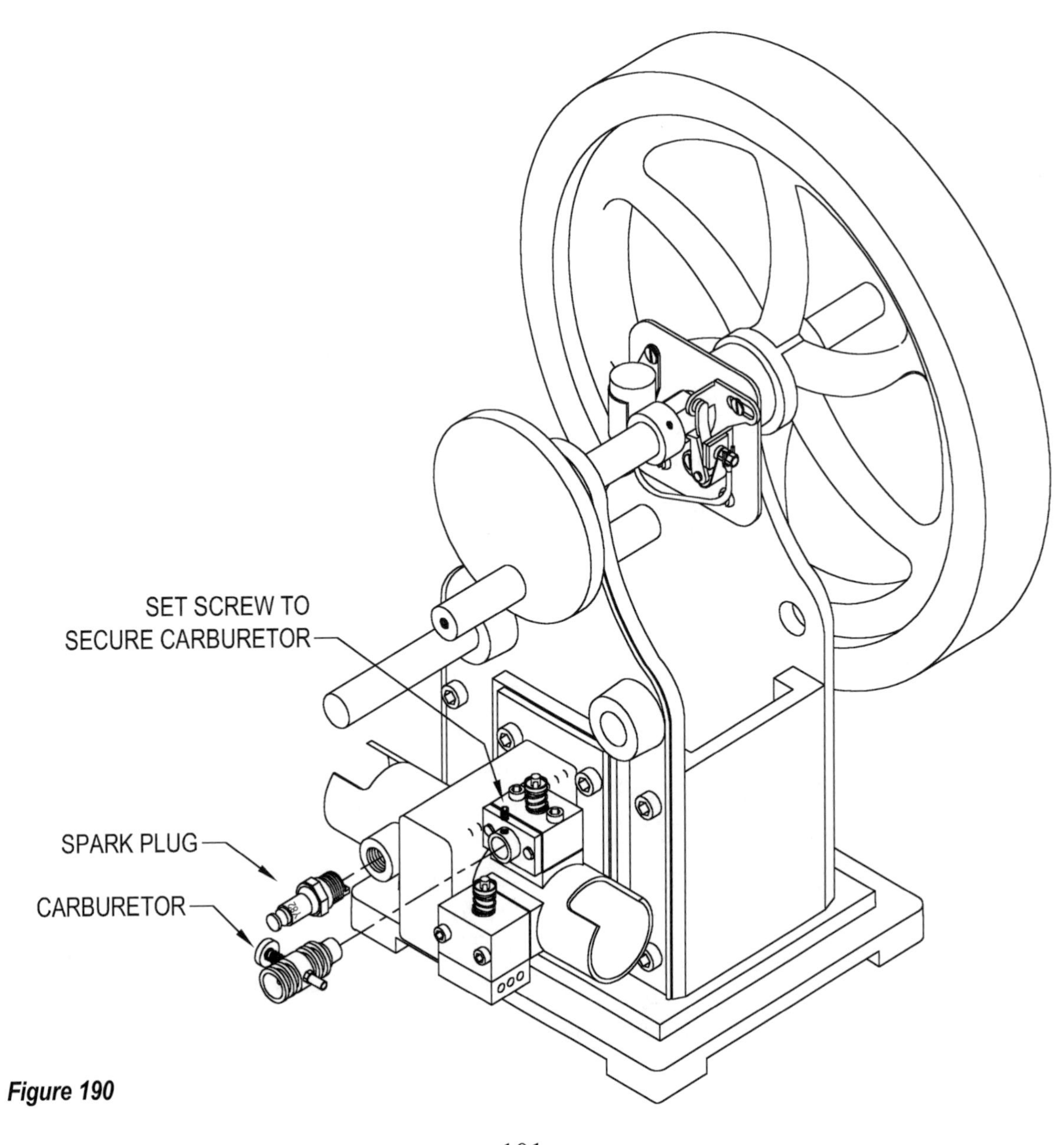
SET SCREW TO
SECURE CARBURETOR
SPARK PLUG
CARBURETOR

Figure 190

A gas tank . . .

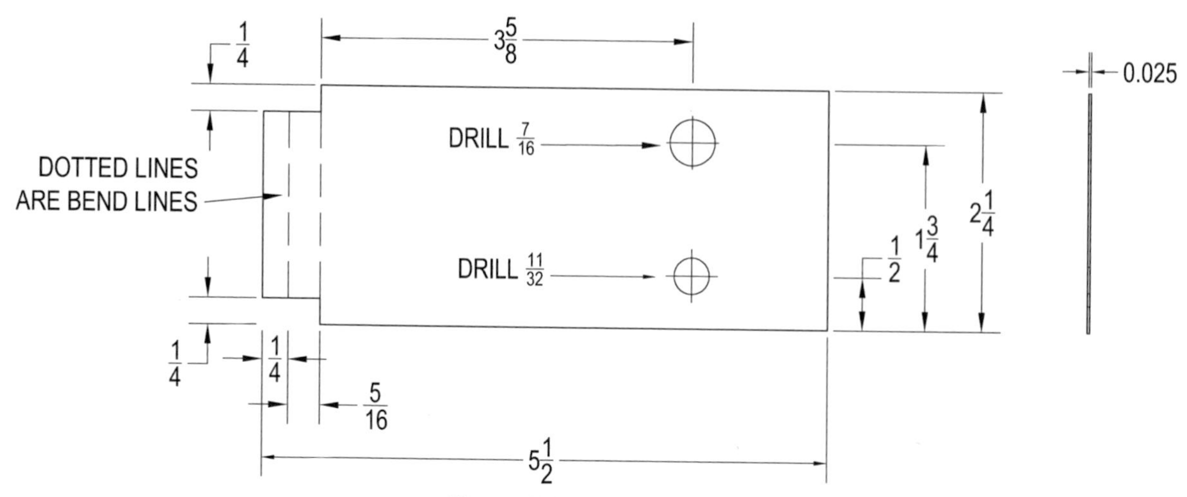

Figure 191 *Gas tank layout*

The gas tank layout is shown in figure 191. We made ours of .025" thick copper sheet, but brass or other sheet metal would work also. The 7/16" hole in the layout is the gas inlet and the 11/32" hole is for the gas outlet. The dotted lines represent the bend marks to form a hook on the tank that will be used to hang the tank from the side of the main frame of the engine. A 5/16" lathe tool bit works well for forming the pocket of the bend.

Figure 193 *The gas tank formed around a 1-7/16" wooden dowel. The form is held in position by a hose clamp on each end. Next step is knock out the wooden dowel and solder the seam.*

To form the tank, roll it over a 1-7/16" diameter wooden dowel. Clamp the ends of formed tank with a hose clamp at each end. Remove the dowel from inside the tank and assemble with solder along the seam.

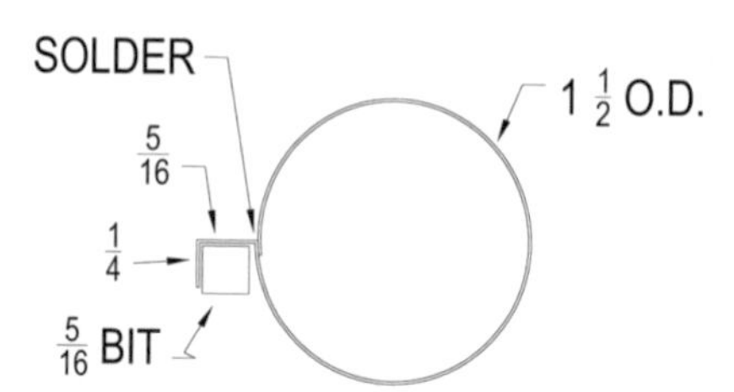

Figure 192 *End view of gas tank*

We devised a die set to form the ends for the gas tank. And with the dies, smooth end caps for the gas tank can be produced from copper, brass or sheet iron in thicknesses from .020" to .032". Dimensions given for the die are for use with 1-1/2" O.D. tubing.

The female die shown in figure 194 contains a tapered recess turned in a block of 2" x 2" mild steel 3/8" thick with the entering edge smoothly rounded. A secondary recess centers the disc. To form the die, mount the

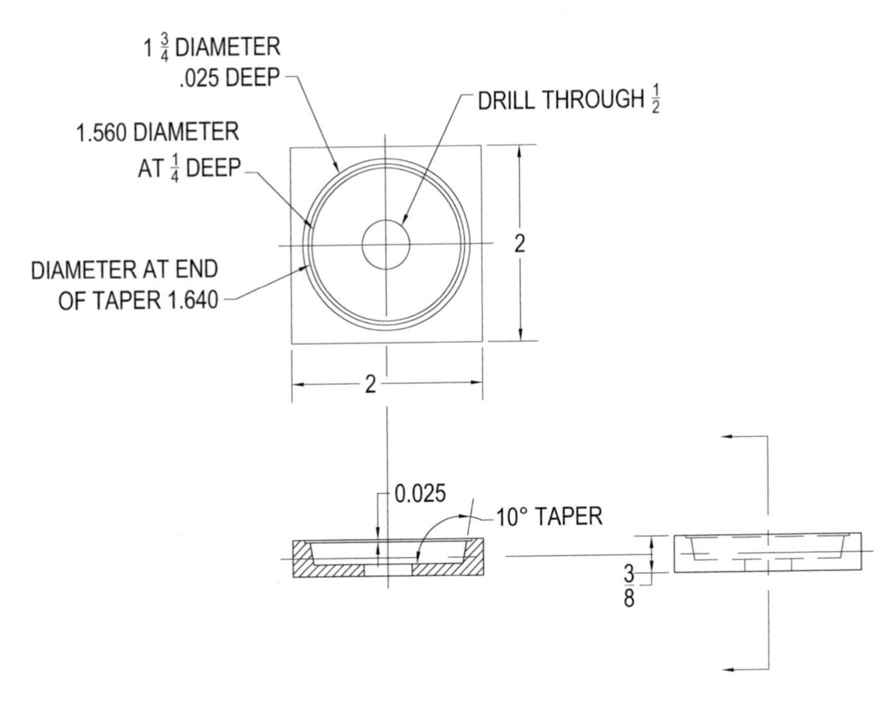

Figure 194 *Female die used to form the ends for the gas tank. Made from 3/8" x 2" x 2" hot or cold rolled steel.*

3/8" x 2" square steel blank in the 4-jaw chuck on the lathe. Bore through the center 1/2" to start the boring bar. Face off smooth and bore the secondary recess about 1/32" deep. Using the boring bar, make a series of facing cuts to rough bore to 1-7/16" diameter 1/4" deep. Set the compound at 10 degrees and proceed to bore to a diameter of 1.560" at the bottom. Leave the edge of the secondary recess sharp at the outside edge, but round at the entering edge. Finish the die with a stone and emery for a smooth surface free of any snag.

One end of the male die shown in figure 195 is turned to 1-7/16" O.D. to force the disc into the female die. It is filed smooth with the edge smoothly rounded.

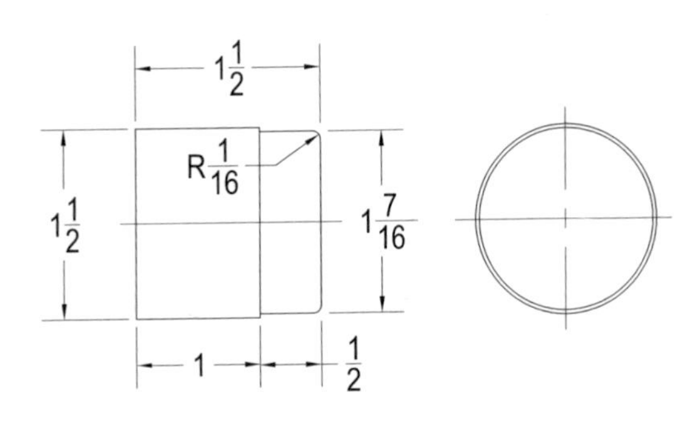

Figure 195 *Male die. Made from 1-1/2" diameter steel round*

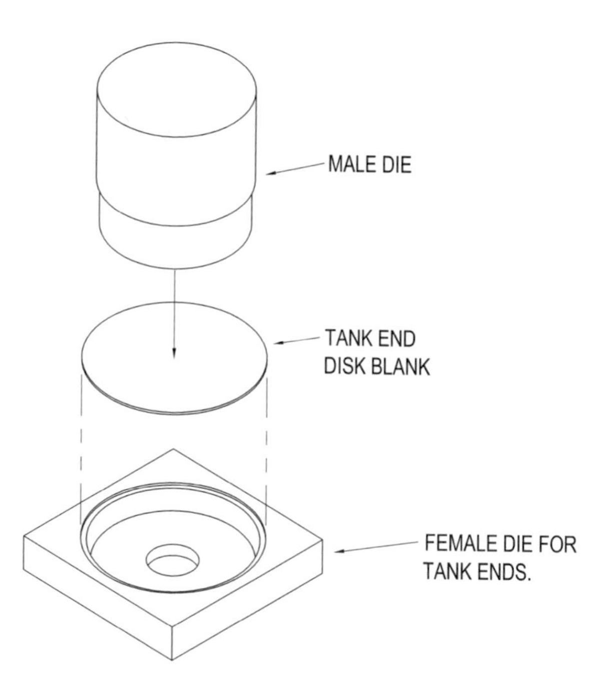

Figure 196 *Set up to form the end cap.*

Figure 197 *The die set up with 2 formed end caps.*

To form the end caps, cut out the 1-3/4" diameter discs to be formed. Oil the disc and set it in the secondary recess. Center the male die on the disc and use the bench vise or other means to press the disc into the female die. Us a dowel through the 1/2" hole to knock the formed end out of the female die.

Solder the formed end caps to the gas tank as shown in figure 198.

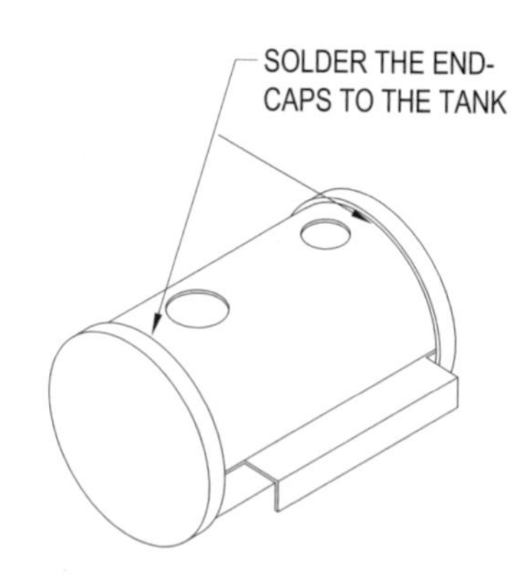

Figure 198 *Solder the end caps to the tank.. IMPORTANT! The inlet and outlet holes in the tank will vent the hot gases from the tank as you solder the ends on. Never solder or weld a sealed tank. Always provide a vent hole for welding gases to escape.*

Gas tank fittings. . .

The tank fittings are as follows. . .

The inlet fitting shown in figure 199. Made from 3/8" diameter brass round then positioned in the 11/32" hole in the tank as shown in figure 201 and soldered in position.

The outlet fitting shown in figure 200. Made from 1/2" brass round then positioned in the 7/16" hole in the tank as shown in figure 201 and soldered in position.

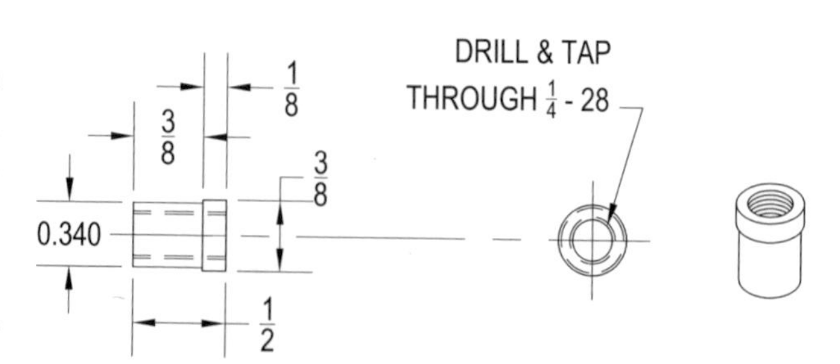

Figure 199 *outlet fitting. Make from 3/8" diameter brass round*

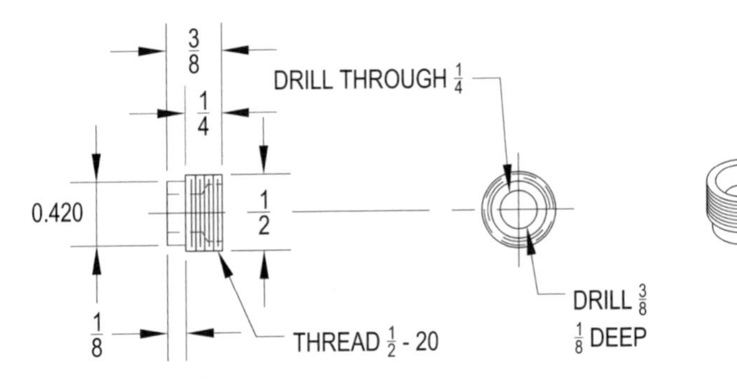

Figure 200 *Gas inlet fitting. Make from 1/2" diameter brass round*

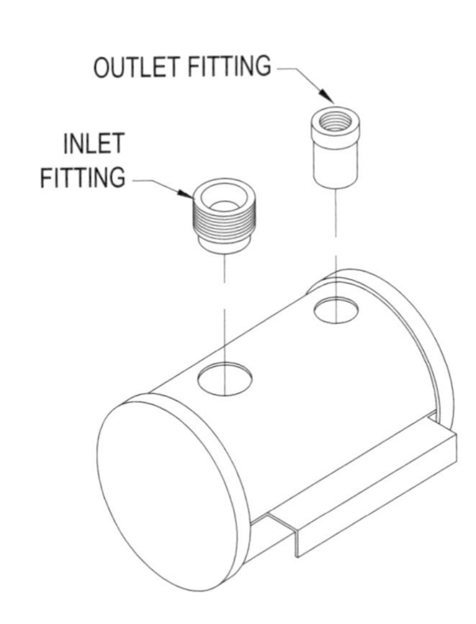

Figure 201 *Insert inlet and outlet fittings in their respective holes and solder in position.*

The gas cap shown in figure 202 and made from 5/8" diameter aluminum or brass round.

The fuel tank adapter made from 3/8" diameter brass round shown in figure 203 and the attached pickup tube also shown in figure 203 and made from 1/8" O.D. brass tubing.

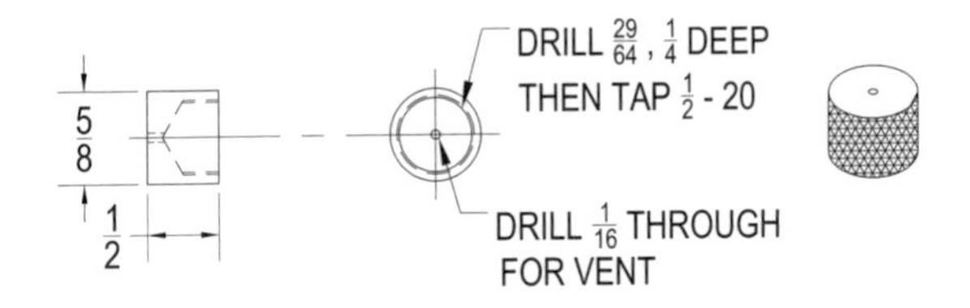

Figure 202 *Gas cap. Make from 5/8" diameter brass or aluminum round.*

And the hose adapter fitting shown in figure 204 made from 1/4" brass hex rod.

The fittings assemble as shown in figure 205. The Check ball is a ball bearing measuring .1565" diameter. Its purpose is to prevent fuel from siphoning out of the gas tank.

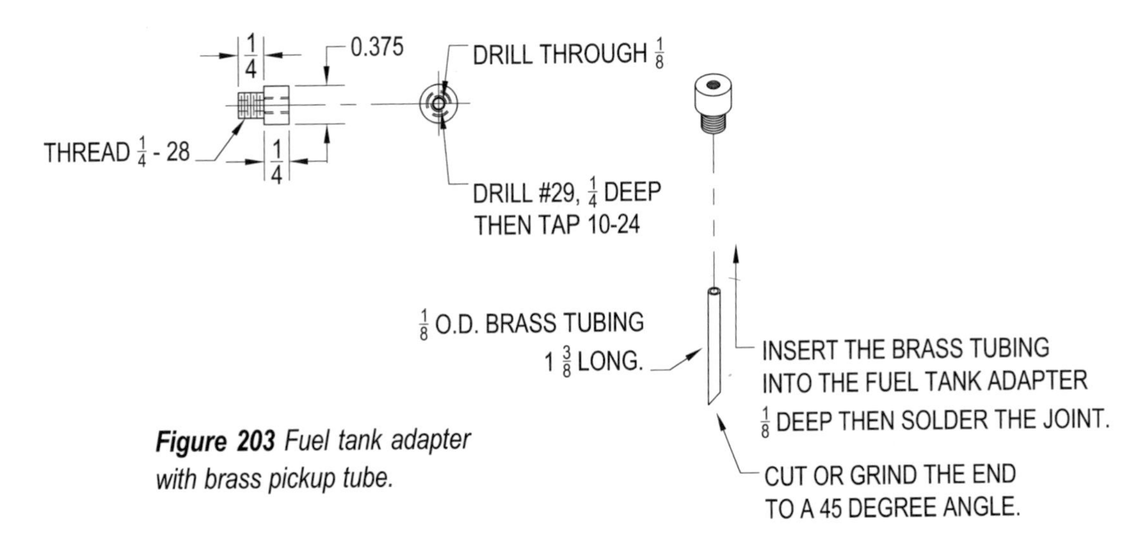

Figure 203 *Fuel tank adapter with brass pickup tube.*

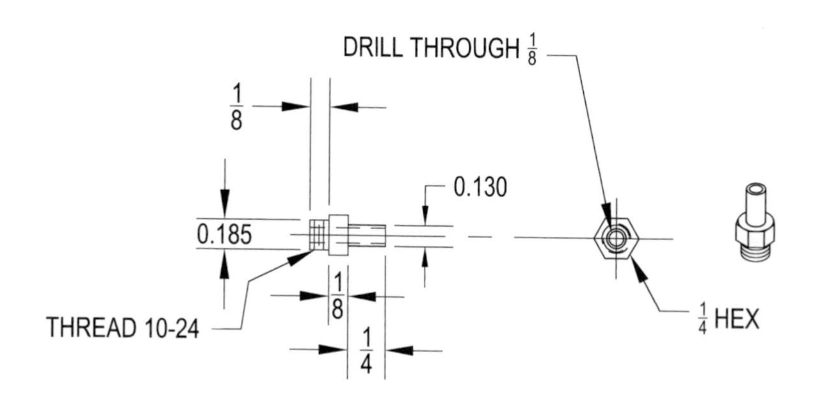

Figure 204 *Hose adapter fitting. Make from 1/4" brass hex.*

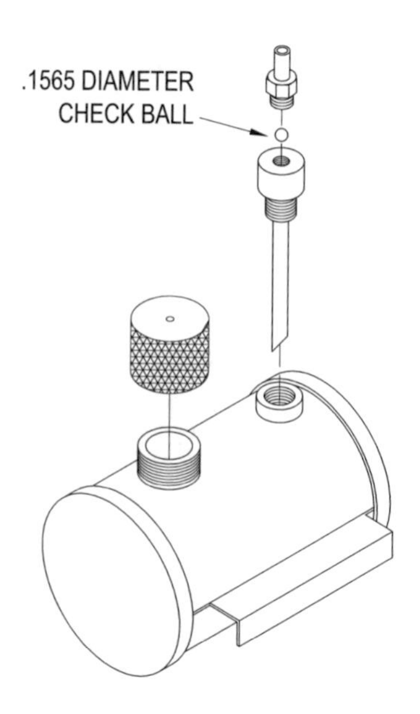

Figure 205

Reassemble the engine . . .

All the parts that you want to paint should now be painted. When that's done, reassemble the engine. The gas tank hangs on the top of the right side panel of the engine main frame. Cut a piece of 1/8" I.D. clear vinyl hose long enough to run between the tank and the carburetor. Connect as shown in figure 206.

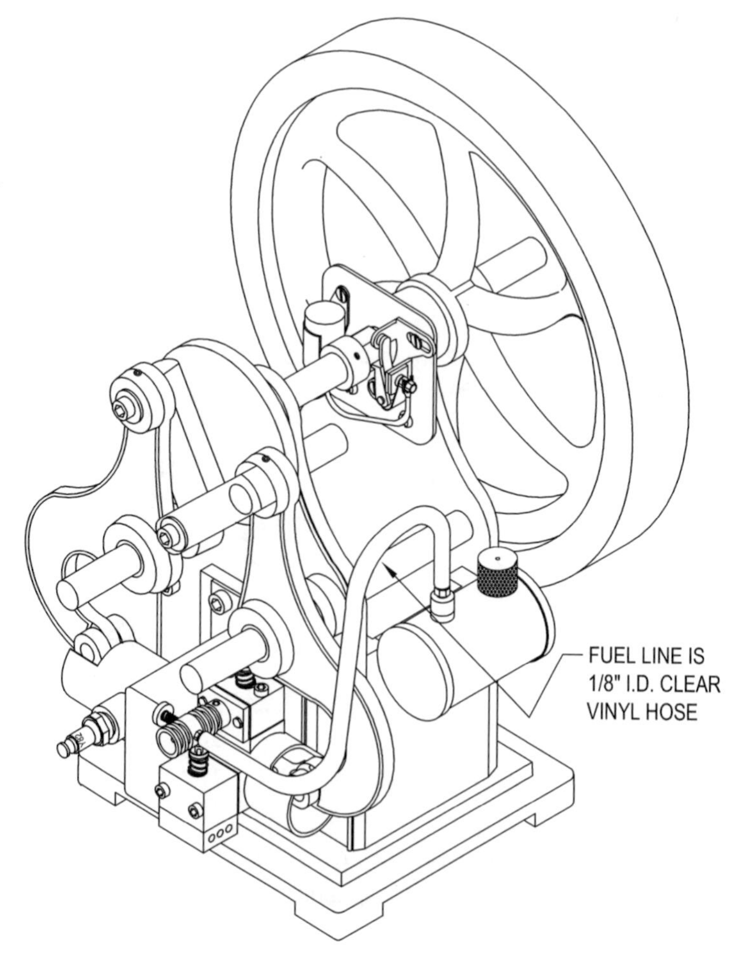

Figure 206 *The reassembled engine with gas tank.*

Add the tie bar . . .

A tie bar is needed to strengthen the oscillating arm pivot shafts. The drawing below gives the details.

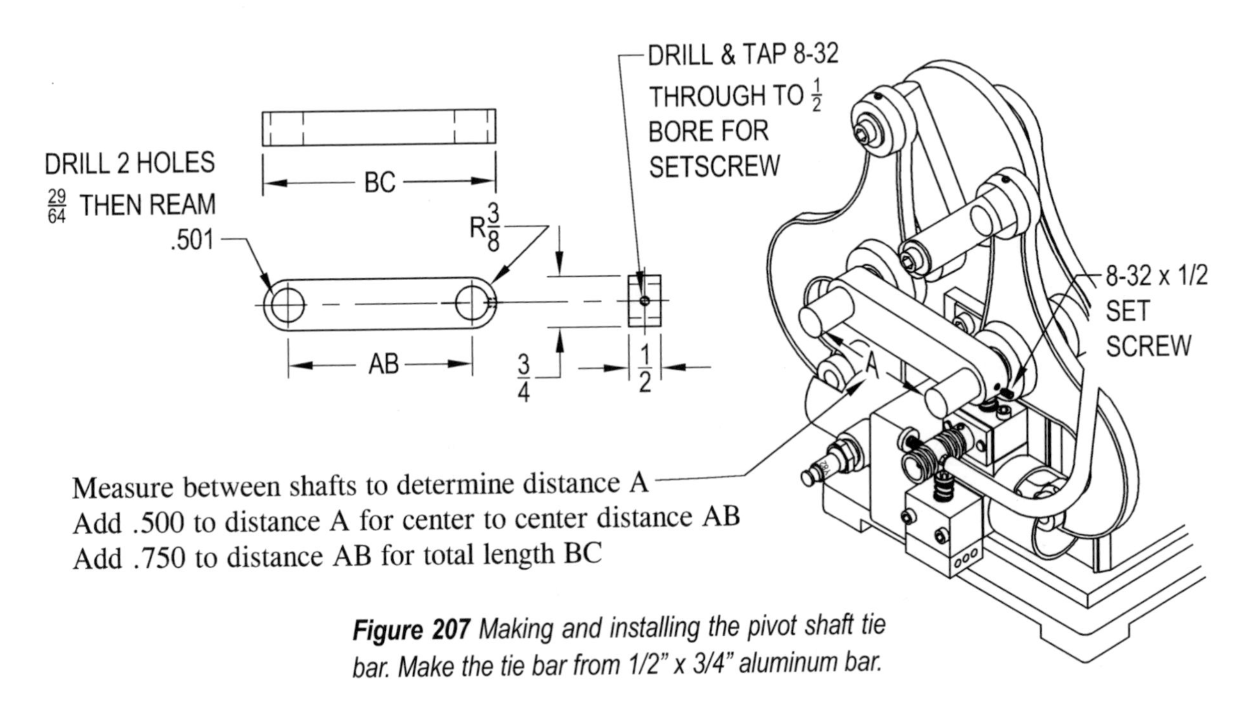

Figure 207 *Making and installing the pivot shaft tie bar. Make the tie bar from 1/2" x 3/4" aluminum bar.*

The electrical system . . .

The electrical charge is provided by a battery and ignition coil. The little lawn mower coils work well for this and the battery can be a small rechargeable battery. The one I used is a 6 volt battery and measures 2-3/4" x 2-3/4" x 3-3/4". Many automotive coils will work also. And with automotive coils you can use a 12 volt battery. Disconnect the battery from the coil when not in use to prevent it from burning out. The coils are expensive if purchased new so check with lawn mower or auto repair shops or salvage yards for used ones. The wiring diagram is shown in figure 208.

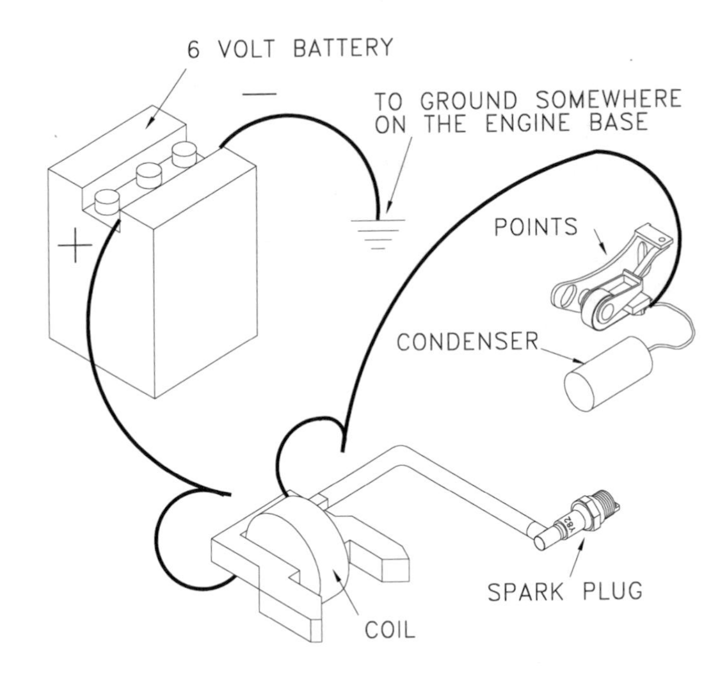

Figure 208 *Wiring diagram*

Setting the ignition cam & Adjusting the points . . .

Timing the engine is simply a matter of adjusting the ignition cam to break the points at about 5 degrees before the piston reaches the top of the compression stroke. Gap the points at around .020. The photo in figure 209 shows the position of the linkage on my engine when the points are set to break.

As a review, you might want to have another look at the photos in figures 153A through 156A on pages 87 & 88. They show the position of the linkage at each of the 4 cycles.

Figure 209 *This is the approximate position of the engine linkage when the points are set to break..*

Final preparation . . .

I wish I could say that all that was required here was to fill the engine with gas, hook up the battery, give the fly wheel a single spin and away she goes. But chances are, that won't be your experience with this engine. Most of the little model engines I have built have required some tinkering and assisted run in time before they would finally overcome friction and develope enough power to run on their own. But this engine requires much more patience than most. The main reasons seem to be compression and friction.

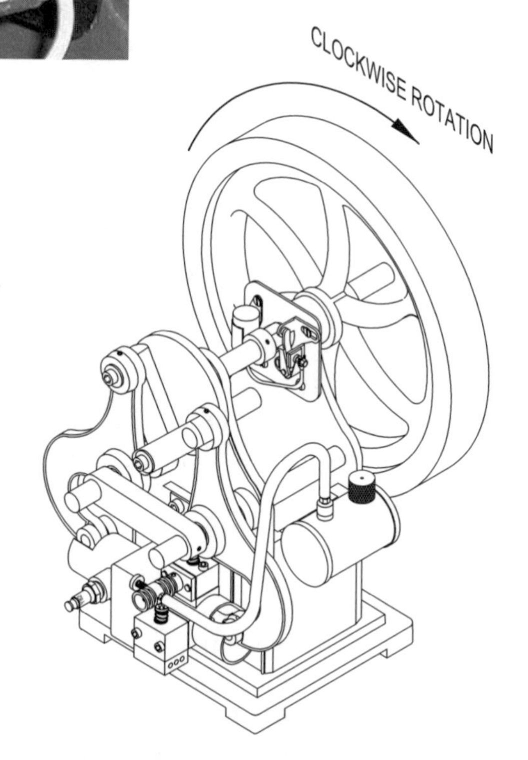

Figure 210 *When the engine is in operation, the flywheel rotates in a clockwise direction as indicated by the arrow.*

In a normal engine, the piston compresses against a solid head and the only chance for compression leaks is either around the piston rings or around the valves. With the Atkinson Differential engine there are 2 pistons in the same cylinder and the pistons compress against each other. This doubles the chance for compression leaks because instead of seating around 1 set of rings you must seat around 2 sets. And with both pistons working in the same cylinder space a slight taper in the cylinder can really cause problems.

And in this engine friction due to the unusual linkage arrangement is another culprit making it difficult to get these engines running.

The solution to both the above problems and an eventual successful run seems to be many hours of assisted run time. The assistance is given by an electric motor belted to the flywheel of the engine. See the photo in figure 211. As you can see in the photo, the engine is bolted or clamped to the work bench.

Figure 211 *Set up we used to break in the engine.*

In the figure, I have attached a board to the bottom of the engine base and then attached the board to the work bench with a couple of screws. The electric motor is also bolted to the work bench and positioned to fit a "V" belt attached to the flywheel. A 1-1/2" pulley is fitted to the shaft of the 1725 rpm electric motor and this arrangement reduces the speed of the engine to around 275 rpm which is just about right. The belt should be a loose fit. Just enough to turn the flywheel without slipping. If it's too tight it will put undue strain on the crank shaft.

The number of hours of assisted run time can only be determined by how well you did on your cylinder and how well your rings fit. On our first 2 running engines, we really did not keep track of the hours of break-in time and tinkering it took to get a running engine, but I can assure you it was many. One thing I have learned on these engine projects is you have to be patient. As I am finishing this book there is a newly constructed engine clacking away in the other room of the shop assisted by an electric motor. I have been running it for about 2 hours every morning for the last few days and so far the total is 22 hours. It's getting close. It has a strong pop when it hits and I remain hopeful, but it still won't run on its own for more than a few seconds. My guess . . . It could take 40 hours or more to break this engine in.

We found it best to supply fuel and spark to the engine during the run in period. The fuel mixture is the same 50:1 oil/gas mixture that you would use in your weed-eater or chainsaw. During the first several hours of break-in time consider mixing an additional 2 ounces per quart of 30 weight non-detergent oil in with the gas. With the added oil it may be possible to get an earlier run on the engine. Then later as the rings seat better you can gradually go back to the straight 50:1 mixture.

Before turning on the electric motor to run the engine in, oil all the moving parts of the engine. Drip some oil on the pistons and squirt a drop or two in the carburetor to lubricate the inside cylinder. Pour antifreeze into the reservoir in the main engine body to a level just slightly higher than the water jacket hole in the front panel. Adjust the carburetor out about 1-1/2 turns and connect the battery and coil to the points and plug. Always make sure the battery is disconnected before adding fuel to the engine. And never add fuel to the engine while it is running.

Now you can turn on the electric motor. The engine will begin clacking away. Lots of moving parts on this engine and those oscillating arms can sure skin a knuckle so watch where you put your hands. Within a few moments you should notice gas being sucked up into the fuel line and beginning its way to the carburetor. If not, choke the end of the carburetor with your finger to bring it on up. Once the fuel reaches the carburetor, adjust the needle valve until you notice the engine starting to fire. It could take an extended run in time before you can get enough compression to get the fuel to ignite, but often you can hear slight pops right away. The pops could be hard to discern, but if you touch your finger to the cylinder at a point above the spark plug and you feel it getting hot then you have ignition.

The spark plug will have a tendency to foul often during the initial break-in period so check it frequently and clean it if necessary. Cleaning the plug is just a matter of shaking out the oil and running a piece of paper between the electrodes to clear the gap. And as with any engine, you might experiment by changing the ignition cam adjustment and carburetor adjustment slightly from time to time to see if performance can be improved.

Keep a close watch on the engine during run in. Keep fuel in the gas tank at all times because the oil/gas mixture is what lubricates the cylinder. A tank of gas will usually last about 45 minutes, but time it to be sure. Then you can do other work in the shop during the run in. When refueling, oil all the moving parts each time you refuel and check for loose screws or linkage.

Disconnect the belt from the flywheel from time to time and give the flywheel a spin by hand to see if the engine will run on its own. When attempting to start the engine it helps to squirt a bit of oil into the end of the carburetor to lubricate the inside cylinder. Choke the engine by placing a finger over the end of the carburetor while giving the flywheel a spin. When it does begin running on its own it will run for a couple of minutes until it gets hot then quit. When it cools down it'll run again for a couple of minutes. Over time it will settle down to be a very reliable running engine. The engines we have running now are one pull starters and run for extended periods of time with little or no problems.

Trouble shooting checklist. . .

Soon after you begin the break in period your engine should start firing enough to heat the cylinder. Below is a list of things to check if you have problems getting ignition.

(1) Is the valve action too stiff? The springs should exert only slight pressure on the valves. Just barely enough to close them.

(2) Are the valves seating properly?

(3) Are the valve bodies fastened tightly to the cylinder? During the initial break in period the bolts fastening the valve bodies have a tendency to work loose.

(4) Is the fuel jet clear? Small particles and lint can clog the jet and restrict fuel flow.

(5) Are the fuel line connections tight?

(6) Are you getting a good spark? Sometimes the battery charge can be low or a loose ground or other connection can cause problems.

(7) Is the spark plug fouled?

(8) Is the ignition cam properly adjusted? It may be necessary to experiment to find optimum adjustment.

My engine is now running on its own power and I thought the notes taken during the run in period for the engine might prove helpful. They are listed below. On each day, I first attempted starting the engine under its own power before resorting to assisted run in.

Day 1. No luck running the engine under its own power. Assisted run in time 1 hour. Ignition was achieved and the cylinder became hot to the touch.

Day 2. No luck running the engine under its own power. Assisted run time 3 hours. No noticeable improvement. During the period, the cylinder was hot to the touch. Lost ignition due to spark plug fouling. Cleared up the problem by cleaning the spark plug.

Day 3. No luck running the engine under its own power. Assisted run time 4 hours. Today was really a repeat of yesterday with the spark plug fouling on at least a couple of occasions..

Day 4. Today the engine turned over a few times under its own power. Assisted run time 2-1/2 hours. Engine seems to be hitting stronger now. Antifreeze mixture is getting very hot which means good ignition.

Day 5. Couldn't get the engine to turn over on its own power. Assisted run time 4 hours. Seem to be losing ground today. Ignition is intermittent. Not sure what the problem is.

Day 6. No luck getting the engine to run under its own power. Assisted run time 5 hours. Spent time looking for the ignition problem. Finally discovered a ball of lint clogging the fuel jet. Cleaned the fuel jet, but still ignition is not like it should be.

Day 7. Assisted run time 2-1/2 hours. Ignition is still intermittent. Not sure what the problem is.

Day 8. Assisted run time 3 hours. Still problems with ignition.

Day 9. Assisted run time 3 hours. No luck running the engine on its own power. Still having problems with ignition.

Day 10. Dad stopped by today and pointed out that the valve springs were too stiff. I shortened the springs and ran the engine for 1 hour. The springs were definitely too tight, but still having ignition problems. The engine won't run on its own power.

Day 10. It occurs to me to check the tightness of the bolts that fasten the valve bodies to the cylinder. They were loose and tightening them solved the ignition problem. Assisted run time 1 hour.

Day 11. This is a good day. The engine starts on its own power and runs for 5 minutes then quits. Can't get it restarted, but dad says this is normal and that it just needs a bit more break in time. Total assisted run time so far is 29 hours. I continued with assisted run time today for 1-1/2 hours.

Day 12. Another good day. The engine runs on its own for about 4 minutes. This time I was able to restart it again for another run of about 5 minutes. No assisted run in time today.

The next several days were repeats of day 12 with longer runs achieved on each successive day. Now, I oil the moving parts of the engine, place a drop of oil on each piston and squirt a drop into the end of the carburetor to lubricate the inside cylinder, give the flywheel a spin and away she goes. And it runs for extended periods of time now or until I tire of watching it.

Final thoughts . . .

Congratulations on the completion of your very own Atkinson Differential Engine. Hopefully you're up and running. If not, you soon will be. One thing for sure, there are not going to be many of these engines around. So if you have completed the project you certainly have something to be proud of. And if you have built the Atkinson Cycle engine you are going to be even more amazed by this one. As I look back on the many hours spent on this project I can't help but continue to admire the ingenious mind that James Atkinson possessed. And this was stuff thought up over 100 years ago. Simply amazing. Anyway, we certainly have had a lot of enjoyment building the Atkinson engines and I hope you have too!

Castings . . .

I may offer for sale a very limited number of casting kits for this engine. These will not be commercial castings, they will be unmachined aluminum castings I pour myself. How many and how long I offer these kits will depend on how long my back holds out. If interested write, call or e-mail for price and availability to:

David J. Gingery Publishing LLC
P.O. Box 318
Rogersville, MO, 65742
Phone 417 890 1965
email gingery@cland.net

Other sources . . .

The following listings are companies that supply tools, materials and books related to the metal working hobby.

Blue Ridge Machinery & Tools
Box 536
Hurricane, WV 25526
Phone 304 562 3538

Camden Miniature Steam
Barrow Farm, Rode,
Bath. BA3 6PS
United Kingdom
Phone 01373 830151

Campbell Tools
2100 Selma Rd.
Springfield, OH 45505
Phone 513 322 8562

Centaur Forge
P.O. Box 340
Burlington, WI 53105
Phone 414 763-9175

Lindsay Publications
P.O. Box 538
Bradley, IL 60915
Phone 815 935-5353
See Lindsay's web site for a complete listing of our other book titles at . . .
http://www.lindsaybks.com

Nolan Supply
P.O. Box 6289
111-115 Leo Ave.
Syracuse, New York 13217
Phone 800 736 2204
http://www.nolansupply.com

Otto Gas Engine Works
2167 Blue Bell Rd.
Elkton, Maryland 21921-3330
Phone 410 398 7340
http://www.dol.net/~dave.reed/otto.html

Plough Book Sales
P.O. Box 14
Belmont, Vic. 3216
Australia
Phone 052 661262

Power Model Supply
13260 Summit Drive
NW Cor Hwy 67 & Long Rd.
Desoto, MO 63020
Phone 314 586-6466

Sulphur Springs Steam Models
P.O. Box 6165
Chesterfield, MO 63006-6165
Phone 314 527-8326

Gas fired furnaces & foundry equipment:

Pyramid Products
85357 American Canal Rd.
Niland, CA 92257
Phone 760 354 4265